Dynamic Programming and Its Application to Optimal Control

This is Volume 81 in
MATHEMATICS IN SCIENCE AND ENGINEERING
A series of monographs and textbooks
Edited by RICHARD BELLMAN, *University of Southern California*

A partial list of the books in this series appears at the end of this volume. A complete listing is available from the Publisher upon request.

# DYNAMIC PROGRAMMING AND ITS APPLICATION TO OPTIMAL CONTROL

*R. Boudarel*

INTERNATIONAL COMPANY FOR INFORMATION TECHNOLOGY
LOUVECIENNES, FRANCE

*J. Delmas*

CENTER FOR STUDY AND RESEARCH IN AUTOMATION
TOULOUSE, FRANCE

*P. Guichet*

INTERNATIONAL COMPANY FOR INFORMATION TECHNOLOGY
LOUVECIENNES, FRANCE

*Translated by R. N. McDonough*

DEPARTMENT OF ELECTRICAL ENGINEERING
UNIVERSITY OF DELAWARE
NEWARK, DELAWARE

ACADEMIC PRESS New York and London 1971

Originally published in the French language under the title
"Commande Optimale des Processus," Tome 3,
"Programmation Dynamique et ses Applications,"
and copyrighted in 1968 by Dunod, Paris.

ACADEMIC PRESS, INC.
111 Fifth Avenue, New York, New York 10003

*United Kingdom Edition published by*
ACADEMIC PRESS, INC. (LONDON) LTD.
Berkeley Square House, London W1X 6BA

LIBRARY OF CONGRESS CATALOG CARD NUMBER: 79-154364

AMS (MOS) 1970 Subject Classification: 49C05

PRINTED IN THE UNITED STATES OF AMERICA

# Contents

## Chapter 9 **Numerical Optimal Control of a Stochastic Process**



# PART 4
# CONTINUOUS PROCESSES

## Chapter 10 **Continuous Deterministic Processes**



## Chapter 11 **Continuous Stochastic Processes**



# PART 5
# APPLICATIONS

## Problem 1 **Introductory Example**



## Problem 2 **Minimum Use of Control Effort in a First-Order System**

## Problem 3 **Optimal Tabulation of Functions**



## Problem 4 **Regulation of Angular Position with Minimization of a Quadratic Criterion**



## Problem 5 **Control of a Stochastic System**



## Problem 6 **Minimum-Time Depth Change of a Submersible Vehicle**



## Problem 7 **Optimal Interception**

## Problem 8 **Control of a Continuous Process**



## Appendix **Filtering**

# Foreword

This book provides a rather complete treatment of optimal control theory for dynamic systems, both linear and nonlinear, discrete and continuous, and deterministic and stochastic, using only simple calculus, matrix notation, and the elements of probability theory. This is accomplished by consistent and exclusive use of the intuitively clear and appealing concepts of dynamic programming in the development of the subject, in contrast to the majority of introductory texts, which follow the more traditional developments based on the calculus of variations. Some rather advanced results of the subject, including those in the stochastic area, are thus made available to readers with a more modest mathematical background than has generally been the case in the past. More so than any previous text, this book illustrates the power and simplicity of the dynamic-programming approach to optimal control problems, and firmly makes the point with solutions to nontrivial example problems, reduced to computational algorithms.

R. N. McDonough

# Preface

This is the third book of a series devoted to optimal process control. In the first volume, the control problem was introduced, and the basic ideas discussed, while in the second volume, optimal control laws were investigated through the mechanism of nonlinear programming [1].

Among all the methods proposed in recent years for determining optimal control laws, dynamic programming, the subject of the present volume, deals with the control problem in the most direct manner. Created by Bellman, dynamic programming rests on the principle of optimality, which is more heuristic than theoretical. Dynamic programming was first successfully applied to problems of economics, but control theorists very quickly recognized in it a powerful theoretical tool which yielded a direct approach to the synthesis of control systems. Even though the rapid development of digital computers has allowed the matter to progress to the stage of concrete applications, there remain to be solved many problems having to do with somewhat complex systems. These difficulties, which are essentially numerical, are also present in the methods studied in the other volumes of this series.

After a discussion of the general philosophy of dynamic programming, the first part of this volume is devoted to discrete deterministic processes, and the practical aspects of the calculation of the optimal control are examined in detail.

In the second part, random (stochastic) processes are introduced, and it is shown that the structure of the solution to the optimal control problem is very much more complicated than in the preceding, deterministic, case.

In the third part, the special case of a discrete linear system, with noise having a rational spectrum, is examined, using a quadratic cost criterion. Many actual systems can be approximated by such a model, and it is shown that the optimal control has a particularly simple structure.

In the fourth part, continuous processes are dealt with. Although theoretical difficulties arise (partial differential equations), it is shown that dynamic programming methods allow such classical theoretical results as the maximum principle to be established rather easily.

In the last part, a number of applied problems are presented, in order that the reader may test his understanding. A short appendix is included, dealing with filtering and prediction, since these play an important role in linear processes with random perturbations present.

We wish to thank all those who have helped us, directly or indirectly, in the preparation of this book, and in particular, MM. Gille, Pelegrin, and Senouillet.

# Dynamic Programming and Its Application to Optimal Control

# PART 1 Discrete Deterministic Processes

# Chapter 1 The Principles of Dynamic Programming

In this short introduction, we shall present the basic ideas of dynamic programming in a very general setting. If the presentation seems somewhat abstract, the applications to be made throughout this book will give the reader a better grasp of the mechanics of the method and of its power.

## 1.1 General Description of the Method

Consider a problem depending on $k$ given numerical quantities $d_1, \ldots, d_k$ and having for solution $m$ quantities $s_1, \ldots, s_m$. Often the direct solution of such a problem is quite difficult. We will use an indirect method, which, although it appears only to increase the complexity of the problem, will be justified by its success.

The procedure evolves in three steps:

1. The given data $d_1, \ldots, d_k$ are considered to be parameters, and the solution to the problem is sought for general values of the parameters. If a solution exists, the values $s_1, \ldots, s_m$ will be functions of the data:

$$s_i = \mathscr{S}_i(d_1, \ldots, d_k).$$

2. For the effective calculation of the functions $\mathscr{S}_i(\ )$, introduced implicitly, relations are sought which they must satisfy. To that end, numerous different problems are considered together, having among them a dependence which allows expressions to be written which express that dependence.

3. The solution of these relations, if the properties have been well chosen, will allow the general solution to be obtained, and, finally, the solution to the particular problem of interest.

This very general approach, that of immersing the initial problem in a larger class of problems, is called invariant embedding. In the case that the relation among the various problems is derived through the principle of optimality, the method is called dynamic programming.

To illustrate the general method, an example other than one using dynamic programming is first considered. We will then return to the principle of optimality before making some remarks about the validity of this approach.

## 1.2 Example of the General Method

Let us consider a body in vertical motion, and satisfying

$$dV/dt = -g - kV^2, \tag{1.1}$$

where $V$ is the vertical velocity, $g$ the acceleration of gravity, and $kV^2$ the air resistance. The maximum altitude $Z_{max}$ attained by the trajectory is to be found, starting from the initial conditions

$$Z = 0, \qquad V = dZ/dt = V_0. \tag{1.2}$$

The three steps in the solution are:

1. The initial conditions $Z$, $V$ are taken to be arbitrary. The altitude difference $Z_{max} - Z$ is a function only of the initial velocity $V$, since the altitude $Z$ does not enter into the equation of motion (1.1). Let $Z_{max} - Z$ be some function $H(V)$, the increase in altitude corresponding to some initial velocity $V$. Once $H(V)$ has been found, the solution to the problem is

$$Z_{max} = Z + H(V). \tag{1.3}$$

2. To find an implicit relation involving $H(V)$, we consider two problems, having initial conditions $(Z_1, V_1)$ and $(Z_2, V_2)$. If we suppose that $Z_1$ and $Z_2$ are nearly the same, the equation of motion (1.1) yields

$$\begin{aligned} Z_2 &\simeq Z_1 + V_1 \, \Delta t, \\ V_2 &\simeq V_1 - (g + kV_1^2) \, \Delta t, \end{aligned} \tag{1.4}$$

where $\Delta t$ is the time required to pass from $Z_1$ to $Z_2$. The relation which must be satisfied by $H(V)$ is

$$H(V_1) = Z_2 - Z_1 + H(V_2). \tag{1.5}$$

This relation corresponds mathematically to the semigroup property of the solution of the differential equation (1.1).

Using (1.4), Eq. (1.5) becomes

$$\begin{aligned} H(V_1) &\simeq V_1 \, \Delta t + H[V_1 - (g + kV_1^2) \, \Delta t] \\ &\simeq V_1 \, \Delta t + H(V_1) - [\mathrm{d}H(V_1)/\mathrm{d}V_1](g + kV_1^2) \, \Delta t, \end{aligned}$$

which becomes, upon passing to the limit of small $\Delta t$,

$$\mathrm{d}H(V)/\mathrm{d}V = V/(g + kV^2). \tag{1.6}$$

3. To obtain $H(V)$, it only remains to solve Eq. (1.6), which yields

$$H(V) = \int_0^V \frac{u}{g + ku^2} \, \mathrm{d}u = \frac{1}{2k} \ln\left(1 + \frac{k}{g} V^2\right). \tag{1.7}$$

## 1.3 Sequential Decision Problems and the Principle of Optimality

The method presented in Section 1.1 is very general. Its success, however, depends essentially on the possibility of relating two problems, and of expressing that relation in a convenient form.

The method of dynamic programming permits systematic solution of optimization problems with sequential decisions. An optimization problem is one in which, for given $d_i$, quantities $x_j$ are to be found such that some criterion involving the $d_i$ and $x_j$ is optimum, taking into account certain constraint conditions. The choice of each $x_j$ corresponds to a decision, which is to be optimal for the particular $d_i$ considered. A sequential decision problem is one in which, if the values of certain $x_j$ are fixed relative to a problem $P_1$, a new problem $P_2$ can be found in which the $x_j$ already fixed play no part, and for which the $d_i$ depend only on the $d_i$ of problem $P_1$ and the decisions already made.

We shall see that the problem of optimal control of a process is of this type, which is the reason for our interest in the problem. A large number of problems in economics and in nonlinear programming can also be put into this form, using an extension of the initial problem (step 1 in Section 1.1).

This problem definition can be put in the condensed form:

$$\text{Problem } P_1 \quad \begin{cases} \text{Data set } D_1 \\ \text{Decision set } X_1 \end{cases}$$

$$\text{Problem } P_2 \quad \begin{cases} \text{Data set } D_2 \\ \text{Decision set } X_2 \end{cases}$$

$$\text{Relations between } P_1 \text{ and } P_2 \quad \begin{cases} X_2 \subset X_1, \quad X' \cup X_2 = X_1 \\ D_1 \xrightarrow{X'} D_2, \end{cases}$$

where $X'$ corresponds to the initial decisions.

The following property can then be stated: Whatever the data $D_1$ of problem $P_1$ and the initial decisions $X'$, if the decision set $X_1$ is optimal, then the decision set $X_2$ must be optimal for problem $P_2$. This property can be demonstrated immediately by contradiction, and is called the principle of optimality.

## 1.4 An Example of Application of the Principle of Optimality

Let us consider again the problem of Section 1.2. If a vertical thrust $F$ with some maximum value $F_{\max}$ is included, the equation of motion becomes

$$dV/dt = -g - kV^2 + F. \tag{1.8}$$

The problem is now to choose a value of $F$ at each instant, so as to maximize the peak altitude of the trajectory.

As the first step, an ensemble of problems is considered, having various initial velocities $V$, and with $\hat{H}(V)$ the increase in altitude corresponding to initial velocity $V$, using the optimal thrust values compatible with the limitation $F \leq F_{\max}$.

As the second step, two problems are considered. The first, $P_1$, corresponds to an initial velocity $V_1$ at the initial time $t_0$, and thrust $F$ applied during the first $\Delta t$ seconds. The second problem, $P_2$, uses initial velocity $V_2$ at the instant $t_0 + \Delta t$, with $V_2 = V_1(t_0 + \Delta t)$, and thrust after the time $t_0 + \Delta t$ which is identical to that used in problem $P_1$ after that time. After time $t_0 + \Delta t$, the trajectories of problems $P_1$ and $P_2$ are thus identical.

In order that $P_2$ be the continuation of $P_1$ past $t_0 + \Delta t$, the system dynamics require

$$\begin{aligned} Z_2 &\simeq Z_1 + V_1\,\Delta t, \\ V_2 &\simeq V_1 + (-g - kV_1^{\,2} + F)\,\Delta t. \end{aligned} \tag{1.9}$$

Under these conditions, the principle of optimality can be written

$$\hat{H}(V_1) = \max_{0 \leq F \leq F_{\max}} [Z_2 - Z_1 + \hat{H}(V_2)], \tag{1.10}$$

since, if the "policy" of thrust decisions made for problem $P_1$ is optimal, it must also provide an optimal policy beginning at $t_0 + \Delta t$ for problem $P_2$, and thus yield the optimal increase in altitude $\hat{H}(V_2)$, by definition of the function $\hat{H}(\,)$. Substituting (1.9) into (1.10), and developing $\hat{H}(V)$ into a series (retaining first-order terms only) yields

$$\begin{aligned}\hat{H}(V_1) &= \max_{0 \leq F \leq F_{\max}} \left[ V_1 \, \Delta t + \hat{H}(V_1) + (-g - kV_1{}^2 + F) \frac{\mathrm{d}\hat{H}(V_1)}{\mathrm{d}V} \Delta t \right] \\ &= V_1 \, \Delta t + \hat{H}(V_1) - (g + kV_1{}^2) \frac{\mathrm{d}\hat{H}(V_1)}{\mathrm{d}V} \cdot \Delta t \\ &\quad + \max_{0 \leq F \leq F_{\max}} \left[ \frac{\mathrm{d}\hat{H}(V_1)}{\mathrm{d}V} F \cdot \Delta t \right],\end{aligned}$$

the last equality following because only the final term depends on $P_1$. Passing to the limit as $\Delta t$ goes to zero in this yields

$$(g + kV_1{}^2) \, \mathrm{d}\hat{H}(V_1)/\mathrm{d}V = V_1 + \max_{0 \leq F \leq F_{\max}} [F \, \mathrm{d}\hat{H}(V_1)/\mathrm{d}V]. \tag{1.11}$$

If it is assumed that $\hat{H}(V)$ is increasing (which will be verified subsequently), $\mathrm{d}\hat{H}(V_1)/\mathrm{d}V$ is positive, and the maximum occurs for $F = F_{\max}$. The optimal policy is thus to use always the maximum thrust available, which is not surprising.

Thus, finally, the principle of optimality has led to the expression for $\hat{H}(V)$,

$$\mathrm{d}\hat{H}(V)/\mathrm{d}V = V/(g + kV^2 - F_{\max}). \tag{1.12}$$

As the third step in the solution, relation (1.12) allows the function $\hat{H}(V)$ itself to be calculated:

$$\hat{H}(V) = \int_0^V \frac{u}{g + ku^2 - F_{\max}} \, du = \frac{1}{2k} \ln\left(1 + \frac{k}{g - F_{\max}} V^2\right). \tag{1.13}$$

In this simple example, the steps followed are precisely the same as in the general method, with the principle of optimality replacing the semigroup property of the first example.

## 1.5 Remarks

In the theoretical plan, objections can be raised about the second step. In effect, it postulates the existence of a solution function, which, in all rigor, would require a preliminary demonstration. The same is true of the third step, for, even if the preceding steps have yielded a relation which must be satisfied by the sought optimal solution, it is possible that it is not sufficient to define a solution, or, on the other hand, that the relation is satisfied by more than one solution. If the relation has a unique solution, it is thus a necessary and sufficient condition. In the other cases, the relation is no longer sufficient, and it is necessary either to compare the various solutions, or to apply additional considerations to define the optimal solution.

Sometimes hypotheses are also made about the domain of definition or the continuity of the solution function. Here again it is necessary to justify them theoretically. In many applications, notably to discrete processes with bounded horizon, the simple structure of the problem obviates these objections. In other applications, notably to continuous processes, the mathematical difficulties raised are important, and the approach used often has only a heuristic character. The method is justified, in such cases, by its success.

In order to approach the application of dynamic programming to process control with a better understanding of the method, the reader should refer to the first problem of the last part of the book. In that exercise, the mechanics of forming and solving the equations are particularly simple.

# Chapter 2 Processes with Bounded Horizon

## 2.1 Definition of a Discrete Process

We shall consider a discrete process in which the state vector at the instant $n$ is denoted $\boldsymbol{x}_n$ ($n$ an integer). If only a finite time interval $0 \leq n \leq N$ is of interest, the process will be said to have bounded horizon. The state vector† obeys an evolution equation

$$\boldsymbol{x}_{n+1} = \boldsymbol{f}(\boldsymbol{x}_n, \boldsymbol{u}_n, n), \qquad \boldsymbol{x} \in \mathbf{R}^p, \quad \boldsymbol{u} \in \mathbf{R}^q, \tag{2.1}$$

where $\boldsymbol{u}_n$ is the control applied at the instant $n$. [With a stepped system the control is also defined between the times $nT$ and $(n+1)T$.] If the function $\boldsymbol{f}(\,)$ does not depend explicitly on $n$, the process is said to be stationary.

In addition, in a realistic problem the controls can not be chosen arbitrarily, but must satisfy certain constraint relations

$$\boldsymbol{u}_n \in \boldsymbol{\Omega}_n \subset \mathbf{R}^q, \tag{2.2}$$

where $\boldsymbol{\Omega}_n$ can depend on $n$ and $\boldsymbol{x}_n$. The case of $\boldsymbol{\Omega}_n$ dependent on $\boldsymbol{x}_n$ is particularly related to dynamic programming.

If the control $\boldsymbol{u}_n$ corresponds to the transition $\boldsymbol{x}_n \to \boldsymbol{x}_{n+1}$, an elementary return $r(\boldsymbol{x}_n, \boldsymbol{u}_n, n)$ is assigned. A return of the type $\rho(\boldsymbol{x}_{n+1}, \boldsymbol{u}_n, n)$ can easily be put in this form by defining

$$r(\boldsymbol{x}_n, \boldsymbol{u}_n, n) = \rho(\boldsymbol{f}(\boldsymbol{x}_n, \boldsymbol{u}_n, n), \boldsymbol{u}_n, n).$$

The process is summarized in the diagram of Fig. 2.1.

† For an explanation of vector and matrix notation, see Boudarel *et al.* [1, Vol. 1], or any of numerous other texts.

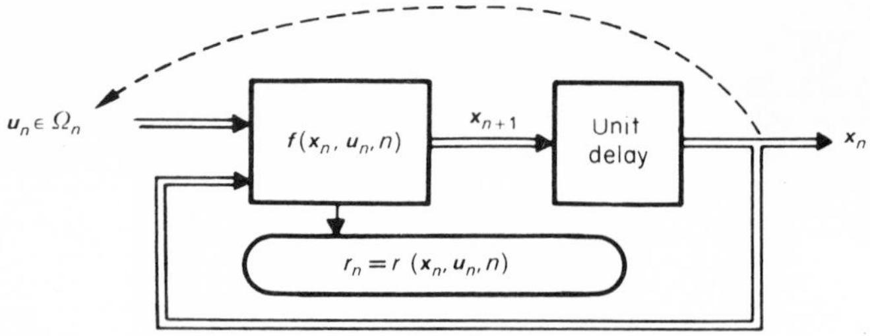


FIG. 2.1. Block diagram of a discrete process.

## 2.2 Statement of the Problem

Starting from an initial state $\boldsymbol{x}_0$, the system is to be carried to a state $\boldsymbol{x}_N$ at the instant $N$, using a sequence of controls $\boldsymbol{u}_0, \ldots, \boldsymbol{u}_{N-1}$, denoted $[\mathrm{U}]_0^{N-1}$, each satisfying the proper constraint relation (admissible controls). The final state $\boldsymbol{x}_N$ may be entirely specified, or may only be required to satisfy some number of terminal constraint relations. In case the final state is entirely unspecified, the system is said to be free.

In general there exist many control policies $[\mathrm{U}]_0^{N-1}$ which carry the system to the desired final state. In this case the final state, or set of states, specified is said to be reachable. A policy is to be found which optimizes the total return

$$R_0 = \sum_{j=0}^{N-1} r_j,$$

which is a function of the initial state $\boldsymbol{x}_0$ and the control policy $[\mathrm{U}]_0^{N-1}$. If the initial time is $n$, the return $\sum_n^{N-1} r_j$ will be denoted $R_n$.

The problem thus stated satisfies the conditions of Section 1.3 and is thus a sequential decision problem. The state and $n$ are the data set $D$, and the controls are the decision set $X$. The principle of optimality may thus be applied.

## 2.3 Application of the Principle of Optimality

The principle of optimality can be stated: Whatever the initial state, if the first decision is contained in an optimal policy, then the remaining decisions must constitute an optimal policy for the problem with initial state the state resulting from the first control decision.

To solve the optimal control problem stated above, the method outlined in Chapter 1 will be used:

1. Rather than the problem $\{\boldsymbol{x}_0, [0, N]\}$, we consider the more general class of problems starting at instant $n$ and state $\boldsymbol{x}_n$:

$$P : \{\boldsymbol{x}_n, [n, N]\}, \qquad 0 \leq n \leq N, \quad \boldsymbol{x}_n \in \mathbf{R}^p.$$

2. The corresponding optimal return $\hat{R}(\boldsymbol{x}_n, n)$ is a function only of $\boldsymbol{x}_n$ and $n$. Since $N$ is fixed, it does not enter as a variable in $\hat{R}$.

3. The principle of optimality then states that, for all $\boldsymbol{x}_n$, if $\boldsymbol{u}_n$ belongs to an optimal policy $[\mathbf{U}]_n^{N-1}$ and results in a state $\boldsymbol{x}_{n+1} = \boldsymbol{f}(\boldsymbol{x}_n, \boldsymbol{u}_n, n)$, then the problem starting at state $\boldsymbol{x}_{n+1}$ must lead to the optimal return $\hat{R}(\boldsymbol{x}_{n+1}, n+1)$. Thus

$$\hat{R}(\boldsymbol{x}_n, n) = \underset{\boldsymbol{u}_n \in \Omega_n}{\text{opt}} \, [r(\boldsymbol{x}_n, \boldsymbol{u}_n, n) + \hat{R}(\boldsymbol{f}(\boldsymbol{x}_n, \boldsymbol{u}_n, n), n+1)]. \tag{2.3}$$

Equation (2.3) is the recurrence equation for the optimal return. It allows successive calculation of $\ldots, \hat{R}(\boldsymbol{x}_n, n), \hat{R}(\boldsymbol{x}_{n-1}, n-1), \ldots, \hat{R}(\boldsymbol{x}_0, 0)$. In case the final state of the system is free, (2.3) is started with $\hat{R}(\boldsymbol{x}_N, N) = 0$. In case terminal constraints are present, the final $k$ controls are not free, since the state $\boldsymbol{x}_{N-k}$ attained determines the controls $\boldsymbol{u}_{N-k}, \ldots, \boldsymbol{u}_{N-1}$. The number $k$ depends on the order of the system and the nature of the constraints on the final state. The return $R(\boldsymbol{x}, N-k)$ is then determined, and used to start the recursion (2.3). We shall return to this point in the following, and it will be illustrated by examples in the final part of the book.

## 2.4 Direct Derivation of the Recurrence Equation

In order to "justify" the principle of optimality, we shall now give a direct derivation of (2.3). By definition,

$$\hat{R}(\boldsymbol{x}_n, n) = \underset{\boldsymbol{u}_n \in \Omega_n, \ldots, \boldsymbol{u}_{N-1} \in \Omega_{N-1}}{\text{opt}} [r(\boldsymbol{x}_n, \boldsymbol{u}_n, n) + \cdots + r(\boldsymbol{x}_{N-1}, \boldsymbol{u}_{N-1}, N-1)].$$

Due to the Markovian nature of the problem, $r(\boldsymbol{x}_n, u_n)$ does not depend on the future controls $\boldsymbol{u}_{n+1}, \ldots$. As a result, the preceding equation can be written

$$\hat{R}(\boldsymbol{x}_n, n) = \underset{\boldsymbol{u}_n \in \Omega_n}{\text{opt}} \{r(\boldsymbol{x}_n, \boldsymbol{u}_n, n) + \underset{\boldsymbol{u}_{n+1} \in \Omega_{n+1}, \ldots}{\text{opt}} [r(\boldsymbol{x}_{n+1}, \boldsymbol{u}_{n+1}, n+1) + \cdots]\}.$$

Recognizing the second term on the right to be the optimal return $\hat{R}(\boldsymbol{x}_{n+1}, n+1)$, this becomes

$$\hat{R}(\boldsymbol{x}_n, n) = \underset{\boldsymbol{u}_n \in \Omega_n}{\text{opt}} \{r(\boldsymbol{x}_n, \boldsymbol{u}_n, n) + \hat{R}(\boldsymbol{x}_{n+1}, n+1)\},$$

which is identical to (2.3). The manipulations involving the operator "optimum" are equivalent to application of the principle of optimality.

## 2.5 Analog Interpretation of the Recurrence Equation

The problem stated above has the analog interpretation indicated by the block diagram of Fig. 2.2 (case of a free system). The element "optimizer" chooses $\boldsymbol{u}_0, \ldots, \boldsymbol{u}_{N-1}$ so as to optimize $\hat{R}(\boldsymbol{x}_0, 0)$. This element

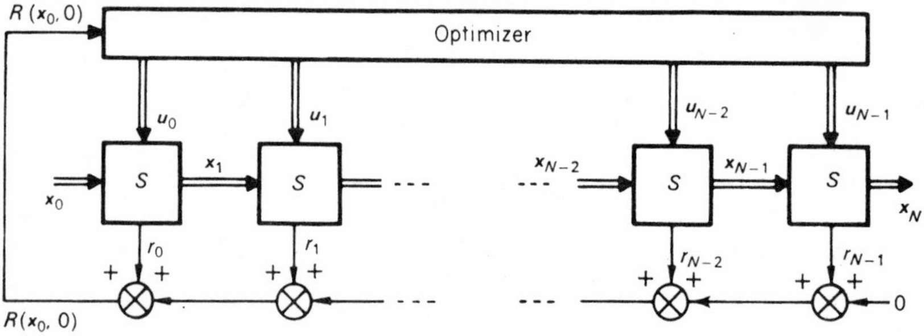


FIG. 2.2. Principle of global optimization.

corresponds to a very complex theoretical problem, and the practical realization of such an element is quite problematic.

The recurrence equation (2.3) allows the scheme of Fig. 2.2 to be simplified to that of Fig. 2.3. The optimizer of Fig. 2.2 has split into a

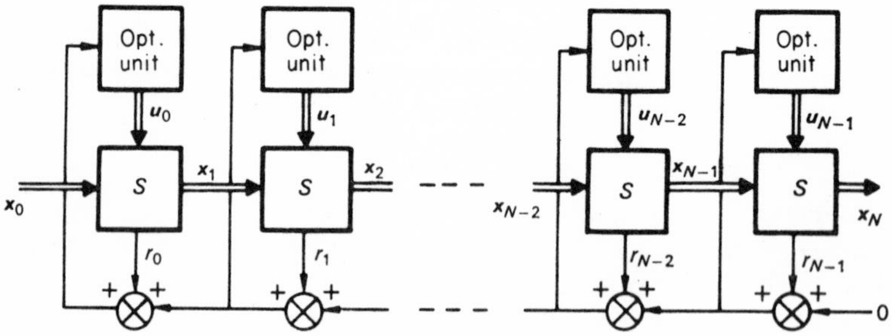


FIG. 2.3 Optimization by dynamic programming.

sequence of elementary optimizers, reducing the general problem to a sequence of identical, and much simpler, problems.

**Remark.** In practice the scheme of Fig. 2.3 provides a method for solution of the problem only in the case of small $N$, since, even aside from the problem of the amount of equipment necessary, the convergence of the elementary loops involves a very complex stability problem.

## 2.6 Practical Application of the Recurrence Equation

Equation (2.3) allows complete solution of the optimal control problem stated above. The solution proceeds in two steps.

### *2.6.1 Calculation of the Optimal Control Functions*

Starting with $\hat{R}(\boldsymbol{x}, N) = 0$ [or with $\hat{R}(\boldsymbol{x}, N - k)$], the $\hat{R}(\boldsymbol{x}, n)$ are calculated successively for decreasing $n$. This can be done by supposing that $\hat{R}(\boldsymbol{x}, n + 1)$ is known either literally or numerically. Then for arbitrary $\boldsymbol{x}$ it is possible to calculate $\hat{R}(\boldsymbol{x}, n)$ using (2.3). If a particular value $\boldsymbol{x} = \boldsymbol{\alpha}$ is considered, then

$$\hat{R}(\boldsymbol{\alpha}, n) = \underset{\boldsymbol{u} \in \Omega_n}{\text{opt}} [r(\boldsymbol{\alpha}, \boldsymbol{u}, n) + \hat{R}(\boldsymbol{f}(\boldsymbol{\alpha}, \boldsymbol{u}, n), n + 1)].$$

The expression in brackets depends only on the variable $\boldsymbol{u}$, and calculation of $\hat{R}(\boldsymbol{\alpha}, n)$ reduces to a search for the optimum of a function of $\boldsymbol{u}$, which is a problem in nonlinear programming (see Boudarel *et al.* [1, Vol. 2], or other texts).

Calculation of $\hat{R}(\boldsymbol{x}, n)$ for each value $\boldsymbol{\alpha}$ of $\boldsymbol{x}$ leads to an optimizing value $\boldsymbol{u}$, and allows point by point construction of the function $\boldsymbol{u}_n = \boldsymbol{g}(\boldsymbol{x}, n)$. These operations are summarized in the flow chart of Fig. 2.4.

Thus solution of the optimal recurrence equation (2.3) permits successive calculation of the optimal returns $\hat{R}(\boldsymbol{x}, n)$, and of the optimal control functions $\boldsymbol{g}(\boldsymbol{x}, n)$.

### *2.6.2 Calculation of the Optimal Control Law*

Use of the control functions $\boldsymbol{g}(\boldsymbol{x}, n)$ allows the solution to be computed without difficulty. It is only necessary to apply the evolution equation of

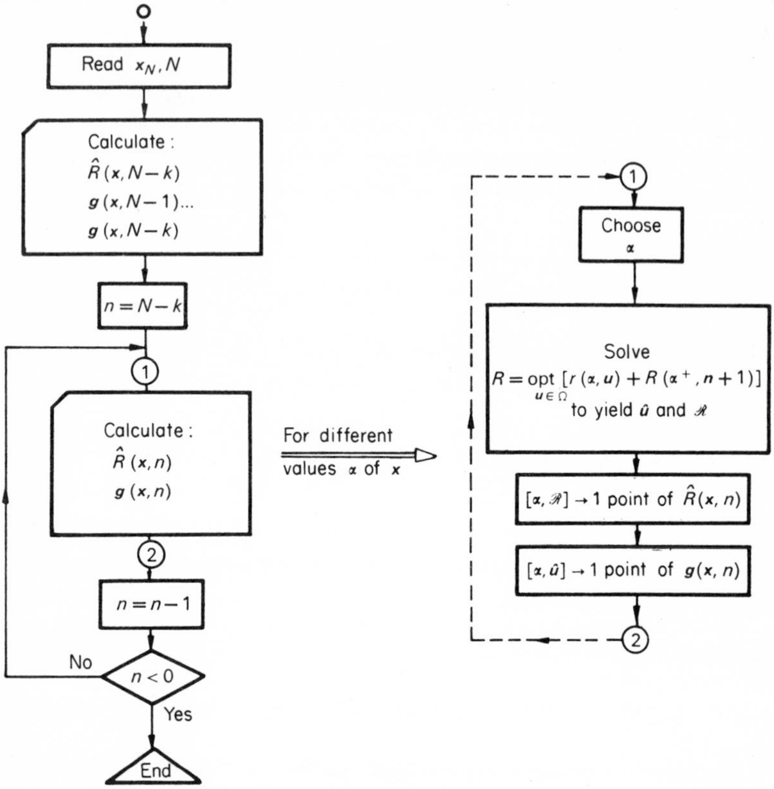


FIG. 2.4. Flow chart for calculation of optimal control functions.

the process successively, using each time the appropriate control function:

$$x_0 \to u_0 = g(x_0, 0) \to x_1 = f(x_0, u_0, 0) \to u_1 = g(x_1, 1), \ldots .$$

This corresponds to the recurrence

$$x_{n+1} = f(x_n, g(x_n, n), n) \tag{2.4}$$

for the optimal trajectory, and can be represented as in Fig. 2.5.

The collection of functions $g(x_n, n)$ allows an optimal controller to be specified which produces the optimal control signal for any state $x_n$. This corresponds to closed-loop control. Usually the controller is realized by a digital computer operating in real time.

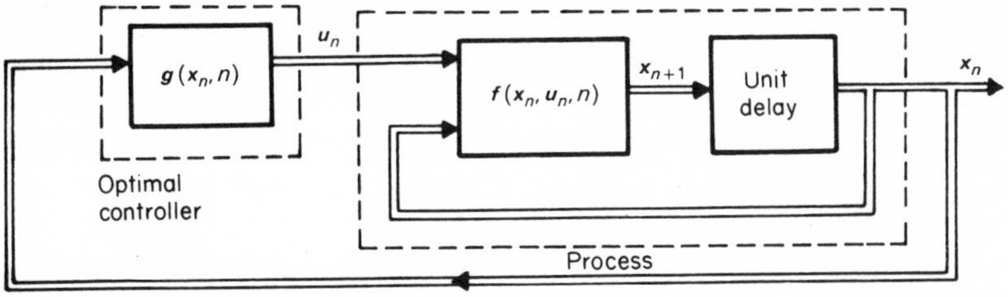


FIG. 2.5. Structure of a controlled process.

**Remark 1.** Calculation of the functions $\boldsymbol{g}(\boldsymbol{x}, n)$, corresponding to synthesis of the controller, is done recursively backward in time, i.e., from time $N - k$ to time 0. On the other hand, the $\boldsymbol{g}(\boldsymbol{x}, n)$ are used in real time in the forward direction, for $n$ from 0 to $N - 1$.

**Remark 2.** The problem considered here was stated for a given $N$. In the case of a nonstationary system, if $N$ is not fixed in advance, the system (2.3) must be solved for each possible $N$. This results in a collection of functions $\boldsymbol{g}(\boldsymbol{x}, N, n)$, corresponding to an optimal return of the type $\hat{R}(\boldsymbol{x}, N, n)$.

On the other hand, if the problem is stationary, $\hat{R}(\boldsymbol{x}, N, n)$ and $\boldsymbol{g}(\boldsymbol{x}, N, n)$ depend only on the combination $N - n$. It is then possible to change the time variable, and compute backward from the terminal time. This can be done by changing from $n$ to $m = N - n$. The return function is then denoted $\hat{R}_m(\boldsymbol{x})$, and the recurrence (2.3) becomes

$$\hat{R}_m(\boldsymbol{x}) = \underset{\boldsymbol{u} \in \Omega}{\text{opt}} \left[ r(\boldsymbol{x}, \boldsymbol{u}) + \hat{R}_{m-1}(\boldsymbol{f}(\boldsymbol{x}, \boldsymbol{u})) \right]. \tag{2.3'}$$

This gives the $\boldsymbol{g}_m(\boldsymbol{x})$ to be used in order of decreasing $m$, as soon as the final time is known. This change of variable is diagrammed in Fig. 2.6.

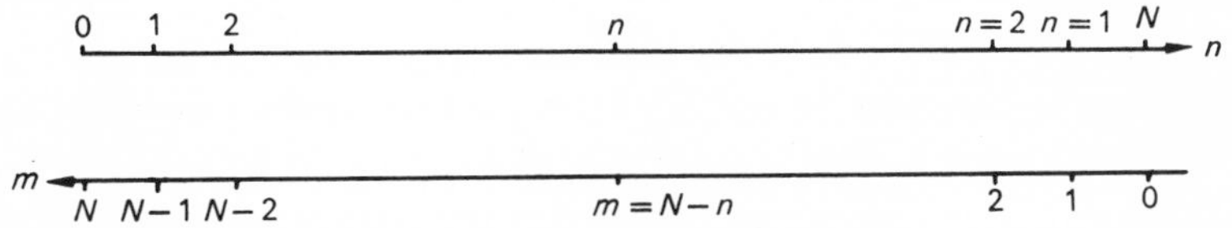


FIG. 2.6. Change of the time variable.

**Remark 3.** Solution by dynamic programming has the advantage that, if the system is stationary, the solution for the problem over the interval $[n - 1, N]$, or over $[n, N - 1]$, can be calculated very rapidly from the solution for $[n, N]$. The optimal recurrence equation need be calculated only one more time to obtain the additional control functions. This is not generally the case with variational methods, which calculate the optimal control law directly relative to a particular horizon. The solution obtained in that case is of no use in solving a problem only slightly different.

**Remark 4.** In dynamic programming methods of solution, it is possible to calculate the optimal return without explicitly calculating the control law. From the practical point of view, this is a considerable advantage. In many cases, realization of the optimal controller using the functions $\boldsymbol{g}_m(\boldsymbol{x})$ involves many difficulties, such as measurement of $\boldsymbol{x}$, calculation and storage of the functions $\boldsymbol{g}_m(\boldsymbol{x})$, behavior of the $\boldsymbol{g}_m(\boldsymbol{x})$ in regard to certain conditions not included in the optimization problem, etc. A simple suboptimal system is thus used as controller (often a linear system). It is, however, important in the choice and justification of the suboptimal controller to compare its performance with that of the optimal controller. The functions $\hat{R}_m(\boldsymbol{x})$ provide the answer to that question. In addition, it is possible to obtain approximate expressions as well for problems of high order (see Chapter 4).

## 2.7 Additive Constraints

In the formulation of the optimal control problem in Section 2.1, only instantaneous constraints were considered, i.e., each $\boldsymbol{u}_n$ was constrained independently. In some problems, global constraints are also present, of the form

$$\mathscr{C} = \sum_{0}^{N-1} C(\boldsymbol{x}_n, \boldsymbol{u}_n, n) \left\{ \begin{matrix} = \\ \leq \end{matrix} \right\} \gamma. \tag{2.5}$$

In the case of an inequality constraint, there exists a domain $\mathscr{D}$ such that if $\boldsymbol{x}_0 \in \mathscr{D}$ the constraint relation is ineffective, while if $x_0 \notin \mathscr{D}$ the constraint is active, and thus must be satisfied as an equality. In the following work, the inequality constraints are either ineffective, or else they reduce to equality constraints. We shall thus consider only the latter type.

### 2.7.1 *Solution Methods*

There are two possible methods for solution of the optimal control problem with global constraints.

#### 2.7.1.1 *Augmentation of the Dimensionality*

The constraint equation (2.5) corresponds to the recurrence relation

$$y_{n+1} = y_n + C(\boldsymbol{x}_n, \boldsymbol{u}_n, n), \tag{2.6}$$

initialized with $y_0 = 0$. Then $\mathscr{C} = y_N$. Adjoining the variable $y$ to the state vector $\boldsymbol{x}$ yields the system

$$\boldsymbol{X}_n = \begin{bmatrix} \boldsymbol{x}_n \\ y_n \end{bmatrix} \to \boldsymbol{X}_{n+1} = \begin{bmatrix} \boldsymbol{f}(\boldsymbol{x}_n, \boldsymbol{u}_n, n) \\ y_n + C(\boldsymbol{x}_n, \boldsymbol{u}_n, n) \end{bmatrix} = \boldsymbol{F}(\boldsymbol{X}_n, \boldsymbol{u}_n, n).$$

With this method, the $p$-dimensional process with sum constraint is transformed to a process of $p + 1$ dimensions with terminal constraints. The methods already discussed are applicable to this augmented problem.

#### 2.7.1.2 *The Use of Lagrange Multipliers*

When the maximum of a function is to be found subject to constraints on the variables, Lagrange multipliers $\lambda_i$ are introduced to construct a modified criterion of the form

$$\text{modified criterion} = \text{criterion} + \sum_i \lambda_i \cdot \text{constraint}_i .$$

The problem is then solved using this modified criterion without taking account of the presence of the constraints. The multipliers $\lambda_i$ are then determined such that the constraints are in fact satisfied.

In our case, this procedure leads to the modified return

$$r'(\boldsymbol{x}_n, \boldsymbol{u}_n, n) = r(\boldsymbol{x}_n, \boldsymbol{u}_n, n) + \lambda C(\boldsymbol{x}_n, \boldsymbol{u}_n, n). \tag{2.7}$$

By solving equation (2.3) for arbitrary $\lambda$, the functions $\hat{R}(\boldsymbol{x}, n, \lambda)$ and $\boldsymbol{g}(\boldsymbol{x}, n, \lambda)$ can be found. A value of $\lambda$ is then found such that (2.5) is satisfied.

### 2.7.2 *Comparison of the Methods*

We shall indicate subsequently that solution of the recurrence equation (2.3) involves computational problems, especially in regard to the amount of memory required, which become increasingly more severe as the dimension of the state vector $\boldsymbol{x}$ increases (the "curse of dimensionality"). The first method intensifies these difficulties. Even though the second method avoids running aground on this particular reef, in principle, its use in practice also encounters difficulties, due this time to the large amount of calculation which is required. This is because the search for $\lambda$ leads to an iterative solution, as shown in the flow chart of Fig. 2.7, if one is to avoid

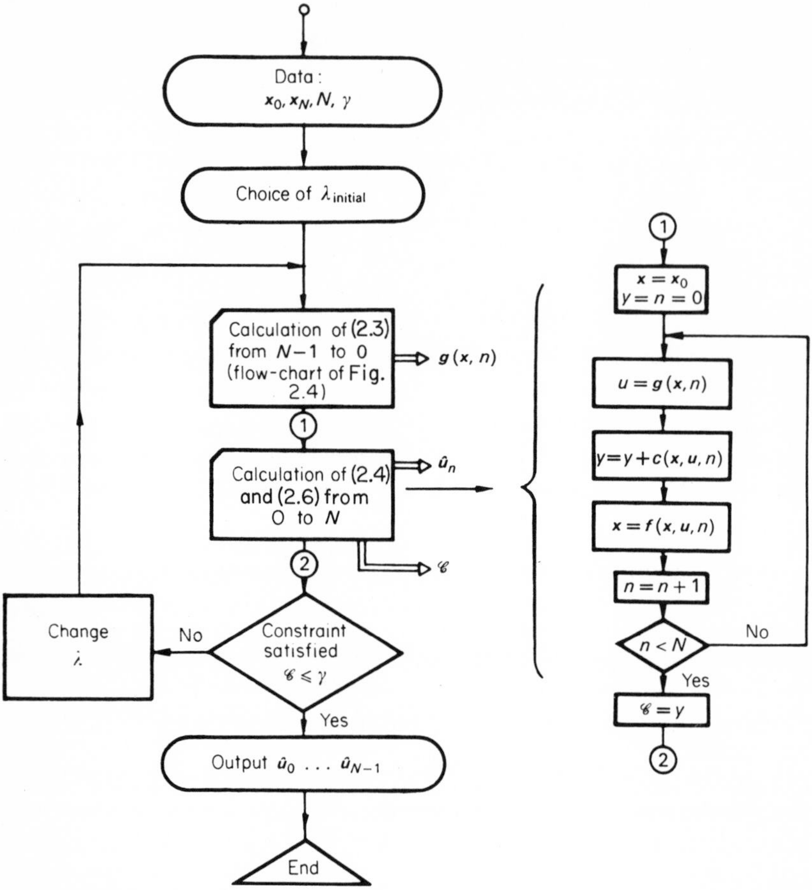


FIG. 2.7. Flow chart for calculation of the optimal control law for a process with an additive-type constraint.

complete calculation of the functions $\hat{R}(\boldsymbol{x}, n, \lambda)$ and $\boldsymbol{g}(\boldsymbol{x}, n, \lambda)$. If these latter functions were calculated in full, the problem would again be of dimension $p + 1$ as in the first method. With this iterative procedure, a direct solution, leading to a closed-loop controller $\boldsymbol{g}'(\boldsymbol{x}, n)$, is no longer possible.

In conclusion, only the first method allows, in theory, solution of the general synthesis problem for a value of the constraint function which is not known in advance, i.e., a closed-loop controller. The second method only allows solution of a particular problem, for which the optimal control can then be calculated, i.e., an open-loop controller. However, by exchanging memory for computation time, it does alleviate the problem of dimensionality.

**Remark 1.** The methods described can, without any theoretical difficulties, be extended to the case of many constraints of additive type.

**Remark 2.** In certain problems it is possible to convert terminal constraints to constraints of the additive type. This decreases the dimensionality of the problem, and a particular solution can be obtained using the second method.

## 2.8 Sensitivity of the Solution

### *2.8.1 General Principle*

Suppose that the mathematical model of the process depends on a parameter $\alpha$ (the development is similar in the case of many parameters). Suppose further that the control functions and optimal return have been calculated using a nominal value $\alpha_0$ of the parameter. It is then reasonable to inquire what is the change in the optimal return when, in the actual process, the parameter has a value $\alpha_0 + \Delta\alpha$ rather than the nominal value $\alpha_0$. This question is particularly important if the given data of a problem are not known precisely. The influence of the parameter $\alpha$ on variations of the criterion can be measured in terms of $\partial\hat{R}/\partial\alpha$, the sensitivity coefficient, or sensitivity function, for variations of $\hat{R}$ with $\alpha$. This is the first term in the series development of $\hat{R}$ as a function of $\alpha$.

### *2.8.2 Calculation of the Sensitivity Coefficient*

Let us suppose that the functions $\boldsymbol{f}(\,)$ and $r(\,)$ are differentiable with respect to $\boldsymbol{x}$, $\boldsymbol{u}$, and $\alpha$. Define

$$\begin{aligned}
\mathbf{F}_x &= \text{the matrix with elements } \partial f_i/\partial x_j\\
\mathbf{F}_u &= \text{the matrix with elements } \partial f_i/\partial u_j\\
\boldsymbol{F}_\alpha &= \text{the vector with elements } \partial f_i/\partial \alpha\\
\boldsymbol{r}_x &= \text{the vector with elements } \partial r/\partial x_i\\
\boldsymbol{r}_u &= \text{the vector with elements } \partial r/\partial u_i\\
r_\alpha &= \partial r/\partial \alpha.
\end{aligned}$$

In addition, let the control function $\boldsymbol{g}(\boldsymbol{x}, n)$ be continuous in $\boldsymbol{x}$. Then a small variation in $\alpha$ results in a variation of the same order in both $\boldsymbol{x}$ and $\boldsymbol{u}$.

Differentiating the return with respect to $\alpha$ yields

$$\frac{\partial \hat{R}}{\partial \alpha} = \frac{\partial}{\partial \alpha}\left(\sum_n r(\boldsymbol{x}_n, \boldsymbol{u}_n, n)\right) = \sum_n \left(\boldsymbol{r}_x{}^{\mathrm{T}} \frac{\partial}{\partial \alpha} \boldsymbol{x}_n + \boldsymbol{r}_u{}^{\mathrm{T}} \frac{\partial}{\partial \alpha} \boldsymbol{u}_n + r_\alpha\right).$$

Let

$$\boldsymbol{y}_n = \partial \boldsymbol{x}_n / \partial \alpha, \qquad \boldsymbol{v}_n = \partial \boldsymbol{u}_n / \partial \alpha.$$

With closed-loop control,

$$\boldsymbol{v}_n = \partial \boldsymbol{g}(\boldsymbol{x}, n)/\partial \alpha = \mathbf{G}_x \boldsymbol{y}_n.$$

Differentiating (2.1) with respect to $\alpha$ yields

$$\boldsymbol{y}_{n+1} = \mathbf{F}_x \boldsymbol{y}_n + \mathbf{F}_u \boldsymbol{v}_n + \boldsymbol{F}_\alpha, \qquad \boldsymbol{y}_0 = \boldsymbol{0}.$$

Solution of this system of recurrence relations allows calculation of the sum defining the sensitivity:

$$\frac{\partial \hat{R}}{\partial \alpha} = \sum_n (\boldsymbol{r}_x{}^{\mathrm{T}} \boldsymbol{y}_n + \boldsymbol{r}_u{}^{\mathrm{T}} \boldsymbol{v}_n + r_\alpha).$$

### 2.8.3 *Sensitivity Constraints*

If the problem is such that the value of the parameter $\alpha$ is not well known, it may be desirable that the optimal return be nearly the same for values of $\alpha$ near $\alpha_0$ as it is for $\alpha_0$ itself. This can be arranged by requiring that the return sensitivity coefficient vanish for $\alpha = \alpha_0$. The return will then be stationary around $\alpha_0$. This condition can be expressed as a constraint of the additive type for the problem involving the total state vector made up of $\boldsymbol{x}_n$ and $\boldsymbol{y}_n$:

$$\text{process} \quad \begin{cases} \boldsymbol{x}_{n+1} = \boldsymbol{f}(\boldsymbol{x}_n, \boldsymbol{u}_n, n), \\ \boldsymbol{y}_{n+1} = \mathbf{F}_x(\boldsymbol{x}_n, \boldsymbol{u}_n, n)\boldsymbol{y}_n + \mathbf{F}_u(\boldsymbol{x}_n, \boldsymbol{u}_n, n)\boldsymbol{v}_n + \boldsymbol{F}_\alpha(\boldsymbol{x}_n, \boldsymbol{u}_n, n), \end{cases}$$

$$\text{return} = \sum_n r_n(\boldsymbol{x}_n, \boldsymbol{u}_n, n),$$

$$\text{constraint} = \sum_n \boldsymbol{r}_x{}^{\mathrm{T}}(\boldsymbol{x}_n, \boldsymbol{u}_n, n)\boldsymbol{y}_n + \boldsymbol{r}_u{}^{\mathrm{T}}(\boldsymbol{x}_n, \boldsymbol{u}_n, n)\boldsymbol{v}_n + r_\alpha(\boldsymbol{x}_n, \boldsymbol{u}_n, n) = 0.$$

The introduction of sensitivity constraints doubles the order of the process, introduces an equality constraint of the additive type, and associates the new quantities $\boldsymbol{v}_n$ with the controls $\boldsymbol{u}_n$.

# Chapter 3 | Processes with Infinite or Unspecified Horizon

## 3.1 Processes with Infinite Horizon

In the previous chapter a process was considered which evolved between the times 0 and $N$. For given $N$, we were able to find an optimal solution. It is reasonable to ask what this solution becomes when $N$ increases and tends toward infinity.

Here we will only consider stationary processes. In this case, if a solution exists, the optimal return for the horizon $[n, \infty]$ depends only on $\boldsymbol{x}$ since $n$ can always be taken as the time origin. The optimal return can thus be written $\hat{R}(\boldsymbol{x})$. The corresponding control function is also stationary, and can be written $\boldsymbol{g}(\boldsymbol{x})$.

Then by application of the principle of optimality

$$\hat{R}(\boldsymbol{x}) = \underset{\boldsymbol{u} \in \Omega}{\text{opt}}[r(\boldsymbol{x}, \boldsymbol{u}) + \hat{R}(\boldsymbol{f}(\boldsymbol{x}, \boldsymbol{u}))]. \tag{3.1}$$

This equation for the optimal return can also be obtained from (2.3) by passing to the limit $n \to \infty$, since then

$$\hat{R}(\boldsymbol{x}, n) \to \hat{R}(\boldsymbol{x}), \qquad \hat{R}(\boldsymbol{x}, n+1) \to \hat{R}(\boldsymbol{x}). \tag{3.2}$$

## 3.2 Processes with Unspecified Horizon

In some processes the final time is not specified in advance, but rather the process stops whenever certain conditions are fulfilled:

$$\boldsymbol{x}_n \in \Phi_n \subset \mathbf{R}^p \Rightarrow n = N, \tag{3.3}$$

where $N$ indicates the final time.

If the subset $\Phi_n$ depends on $n$, the problem is one of pursuit. If $\Phi$ is constant, the problem is to steer the system to the given subset. It is also possible to consider problems in which $\Phi_n$ depends on $x_{n-1}$:

$$x_{n+1} \in \Phi(x_n) \Rightarrow n + 1 = N.$$

If the process is stationary, and if $\Phi$ does not depend explicitly on time, the optimal return $\hat{R}(x)$ and the optimal control function $g(x)$ are also stationary.

In such problems, $\hat{R}(x)$ obeys a recurrence relation which can be written directly from the principle of optimality:

$$\hat{R}(x) = \underset{u \in \Omega}{\text{opt}} \, [r(x, u) + \hat{R}(f(x, u))], \qquad x \notin \Phi \tag{3.4}$$

$$\hat{R}(x) \equiv 0, \qquad x \in \Phi. \tag{3.5}$$

Equations (3.1) and (3.4) differ only in the presence of the boundary condition (3.5).

The problem of steering to the origin in minimum time is an example of the use of (3.4) and (3.5), in which $r(x, u) = 1$ and $\Phi = \{0\}$. Letting the attained minimum time be $\hat{T}(x)$, (3.4) and (3.5) become

$$\hat{T}(x) = \min_{u \in \Omega} \, [1 + \hat{T}(f(x, u))],$$

$$\hat{T}(0) \equiv 0.$$

## 3.3 Structure and Stability of a System with Infinite Horizon

In the preceding two sections, it was shown that the optimal solution, if it exists, is obtained from a stationary control function $g(x)$. This corresponds to a closed-loop structure analogous to that in Fig. 2.5. It can then be asked how the free closed-loop system behaves as $n \to \infty$, i.e., when its stability can be investigated. This concept has been discussed by Boudarel *et al.* [1, Vol. 1] as summarized in the following.

A free stationary system is said to be asymptotically stable if, for any initial state $x_0$, the corresponding trajectory converges toward a point $A$, called a fixed point, such that, if the system equation is $x_{n+1} = \Phi(x_n)$, the point $A$ satisfies $A = \Phi(A)$. If in addition the trajectories are continuous with respect to the initial point, the system is said to be uniformly asymptotically stable.

In investigating the stability of a system, we shall use the following theorem, corresponding to the second method of Lyapunov: Consider the system $\boldsymbol{x}_{n+1} = \boldsymbol{\Phi}(\boldsymbol{x}_n)$, and let $\boldsymbol{A}$ be a fixed point. The system is uniformly asymptotically stable if there exists a scalar function $V(\boldsymbol{x})$, called a Lyapunov function, such that

$$\begin{aligned} V(\boldsymbol{x}) &> 0, & \boldsymbol{x} &\neq \boldsymbol{A}, \\ V(\boldsymbol{x}) &= 0, & \boldsymbol{x} &= \boldsymbol{A}, \\ V(\boldsymbol{x}) &\to \infty, & \|\boldsymbol{x}\| &\to \infty, \\ V(\boldsymbol{x}_{n+1}) - V(\boldsymbol{x}_n) = \Delta V(\boldsymbol{x}_n) &< 0, & \boldsymbol{x}_n &\neq \boldsymbol{A}. \end{aligned}$$

In order to apply this result to our problem, we shall make the following restricting assumptions: (a) The evolution equation of the process is such that $\boldsymbol{f}(\boldsymbol{0}, \boldsymbol{0}) = \boldsymbol{0}$, i.e., the zero state is an equilibrium state. (b) The elementary return satisfies

$$\begin{aligned} &r(\boldsymbol{x}, \boldsymbol{u}) > 0, && \boldsymbol{x} \neq \boldsymbol{0}, \quad \boldsymbol{u} \neq \boldsymbol{0}, \\ &r(\boldsymbol{0}, \boldsymbol{0}) = 0, && \\ &r(\boldsymbol{x}, \boldsymbol{u}) \to \infty, && \|\boldsymbol{x}\| \to \infty, \quad \forall \boldsymbol{u}. \end{aligned}$$

(c) A minimum of the criterion is sought. It is then easy to see that the optimal control law, for $\boldsymbol{x}_0 = \boldsymbol{0}$, is $\boldsymbol{u}_n \equiv \boldsymbol{0}$, leading to a return $\hat{R}(\boldsymbol{0}) = 0$. Thus $\boldsymbol{x} = \boldsymbol{0}$ is a fixed point of the closed-loop system. It can then be verified that $\hat{R}(\boldsymbol{x})$ is a Lyapunov function. In fact,

$$\begin{aligned} &\hat{R}(\boldsymbol{x}) = \sum_0^\infty r(\boldsymbol{x}_n, \boldsymbol{u}_n) > 0, \qquad \boldsymbol{x} \neq \boldsymbol{0}, \\ &\hat{R}(\boldsymbol{0}) = 0, \\ &\hat{R}(\boldsymbol{x}) \to \infty, \qquad \|\boldsymbol{x}\| \to \infty, \\ &\Delta\hat{R}(\boldsymbol{x}) = \hat{R}(\boldsymbol{x}_{n+1}) - \hat{R}(\boldsymbol{x}_n) = \hat{R}(\boldsymbol{x}_{n+1}) - [r(\boldsymbol{x}_n, \boldsymbol{u}_n) + \hat{R}(\boldsymbol{x}_{n+1})] \\ &\qquad = -r(\boldsymbol{x}_n, \boldsymbol{u}_n) < 0, \qquad \forall \boldsymbol{x}_n \neq \boldsymbol{0}. \end{aligned}$$

Thus, under hypotheses (a)–(c), the system is uniformly asymptotically stable.

**Remark 1.** Stability has been established for the fixed point $\boldsymbol{0}$. However, if hypotheses (a)–(c) had been stated for different values of $\boldsymbol{x}$ and $\boldsymbol{u}$, a simple change of variables would have led to the above case.

**Remark 2.** Assumptions (a)–(c) about the process and the return function are in fact satisfied for the majority of practical problems. They simply state that the return is positive everywhere except at the equilibrium point $\boldsymbol{x} = \boldsymbol{0}$. As a consequence, $\hat{R}(\boldsymbol{x}_n)$ is a strictly decreasing sequence which vanishes only at $\boldsymbol{x} = \boldsymbol{0}$. This ensures convergence of the sequence to zero.

## 3.4 Calculation of the Solution of the Optimality Equation

### *3.4.1 Principle of the Method*

The problems discussed in the first two sections both lead to an implicit functional equation for the optimal return. This equation is of a novel type, involving, as it does, the optimization operator. In general, a literal solution is impossible, and an iterative method must be used (Picard's method).

In the iterative method, a sequence of functions $R_n(\boldsymbol{x})$ is considered which satisfies the recurrence relation

$$R_{n+1}(\boldsymbol{x}) = \underset{\boldsymbol{u} \in \Omega}{\text{opt}} \, [r(\boldsymbol{x}, \boldsymbol{u}) + R_n(\boldsymbol{f}(\boldsymbol{x}, \boldsymbol{u}))]. \tag{3.6}$$

If the horizon is not specified in advance, it is required in addition that

$$R_n(\boldsymbol{x}) \equiv 0, \qquad \boldsymbol{x} \in \Phi. \tag{3.7}$$

Note that if for all $\boldsymbol{x} \in \Phi$, there exists a $\boldsymbol{u} \in \Omega$ such that $r(\boldsymbol{x}, \boldsymbol{u}) = 0$ and $\boldsymbol{f}(\boldsymbol{x}, \boldsymbol{u}) \in \Phi$, it is sufficient that condition (3.7) be satisfied for $n = 0$ in order that it be satisfied for all $n$. If the sequence $R_n(\boldsymbol{x})$ converges, it converges toward $R(\boldsymbol{x})$, and $\boldsymbol{g}_n(\boldsymbol{x})$ converges to $\boldsymbol{g}(\boldsymbol{x})$.

In both cases a solution is sought to a functional equation of the type

$$\varphi(\boldsymbol{x}) = \mathcal{F}[\varphi(\boldsymbol{x})], \tag{3.8}$$

using a sequence constructed through the relation

$$\varphi_{n+1}(\boldsymbol{x}) = \mathcal{F}[\varphi_n(\boldsymbol{x})]. \tag{3.9}$$

It should be noted that if an error is made in the calculation of $\varphi_{n+1}(\boldsymbol{x})$ in (3.9), this does not affect the value of the limit toward which the sequence converges, and thus the accuracy of the solution to (3.8) is unaffected [this is on condition that (3.9) converges uniformly to a unique limit].

### *3.4.2 Validity of the Method*

The success of an iterative method such as this depends on the existence of a solution $\varphi(\boldsymbol{x})$ of (3.8), and on the properties of the functional transformation $\mathscr{F}$. The Banach contraction theorem provides a justification of the method [29]: Let $\mathscr{Y}$ be a complete metric space, and let $A$ be an operator over $\mathscr{Y}$ transforming an element $y$ of $\mathscr{Y}$ into an element $A(y)$ of $\mathscr{Y}$. If $d[A(y_1), A(y_2)] \leq a d[y_1, y_2]$, with $a < 1$, where $d[y_1, y_2]$ is the distance between the elements $y_1$ and $y_2$, then the equation $A(y) = y$ has a unique solution $y$, and the sequence $y_n$ defined by $y_{n+1} = A(y_n)$ converges uniformly to y.

Thus, when the contraction property can be established for an operator $\mathscr{F}[\ ]$, the iterative method converges for any initial function $\varphi_0(x)$. This theoretical method is, however, very difficult to apply. Nonetheless, Bellman [2] was able to establish the contraction property for the particular case:

$$\|\boldsymbol{f}(\boldsymbol{x}, \boldsymbol{u})\| < a \|\boldsymbol{x}\|, \qquad a < 1, \quad \forall \boldsymbol{x}, \boldsymbol{u};$$

$$r(\boldsymbol{0}, \boldsymbol{u}) = 0, \qquad \forall \boldsymbol{u}; \qquad r(\boldsymbol{x}, \boldsymbol{u}) > 0.$$

It should be noted that the first condition requires that the system tend to zero regardless of the control applied, so that only the origin is reachable at infinity. From the point of view of applications, this condition is quite restrictive.

### *3.4.3 Initialization*

#### *3.4.3.1 Initialization in Return Space*

The use of Picard's method, when the iteration converges, will be the more efficient as the initial function $\varphi_0(\boldsymbol{x})$ approaches $\hat{R}(\boldsymbol{x})$. Unfortunately, it is difficult to determine the form of $\hat{R}(\boldsymbol{x})$. One method is to choose $\varphi_0(\boldsymbol{x}) = 0$. In the case of a process with infinite horizon, the $\varphi_n(\boldsymbol{x})$ will be the optimal returns corresponding, successively, to free processes with bounded horizons, $[0, n]$. Thus, if the limit solution exists, the sequence $\varphi_n(\boldsymbol{x})$ necessarily converges to $\hat{R}(\boldsymbol{x})$, and the sequence $\boldsymbol{g}_n(\boldsymbol{x})$ to $\boldsymbol{g}(\boldsymbol{x})$.

#### 3.4.3.2 *Initialization in Policy Space*

If it is difficult to predict the form of $\hat{R}(\boldsymbol{x})$, it is often easier to determine that of the optimal function $\boldsymbol{g}(\boldsymbol{x})$, which is of more interest in practice. To an estimate $\tilde{\boldsymbol{g}}(\boldsymbol{x})$ of $\boldsymbol{g}(\boldsymbol{x})$ there corresponds, for the closed system, a return $\tilde{R}(\boldsymbol{x})$ given by

$$\tilde{R}(\boldsymbol{x}) = r(\boldsymbol{x}, \tilde{\boldsymbol{g}}(\boldsymbol{x})) + \tilde{R}(\boldsymbol{f}(\boldsymbol{x}, \tilde{\boldsymbol{g}}(\boldsymbol{x}))). \tag{3.10}$$

This is an implicit equation, which can be solved by Picard's method. Since the operator opt[ ] is not involved, the solution proceeds rapidly.

This implies, however, that $\tilde{R}(\boldsymbol{x})$ exists, that is, that $\tilde{\boldsymbol{g}}(\boldsymbol{x})$ applied to the process leads to a bounded total return, for any $\boldsymbol{x}_0$. This requires in general that $\boldsymbol{x}^+ = \boldsymbol{f}(\boldsymbol{x}, \tilde{\boldsymbol{g}}(\boldsymbol{x}))$ be stable. In addition it is necessary that (3.10) have a unique solution. If $\tilde{R}(\boldsymbol{x})$ exists, the policy $\tilde{\boldsymbol{g}}(\boldsymbol{x})$ is said to be admissible. In any event, if $\tilde{R}(\boldsymbol{x})$ exists, it can always be calculated as the limit of the sum

$$\hat{R}(\boldsymbol{x}) = \sum_{n=0}^{\infty} r(\boldsymbol{x}_n, \tilde{\boldsymbol{g}}(\boldsymbol{x}_n)), \quad \text{with} \quad \boldsymbol{x}_{n+1} = \boldsymbol{f}(\boldsymbol{x}_n, \tilde{\boldsymbol{g}}(\boldsymbol{x}_n)), \quad \boldsymbol{x}_0 = \boldsymbol{x}. \tag{3.11}$$

If $\tilde{\boldsymbol{g}}(\boldsymbol{x})$ has been well chosen, the resulting return $\tilde{R}(\boldsymbol{x})$ will be near $\hat{R}(\boldsymbol{x})$, and if this $\tilde{R}(\boldsymbol{x})$ is taken as $\varphi_0(\boldsymbol{x})$, the sequence will converge rapidly toward the sought $\hat{R}(\boldsymbol{x})$.

### 3.4.4 *Iteration in Policy Space*

Suppose $\boldsymbol{g}^0(\boldsymbol{x})$ is an initial policy, and that for any $\boldsymbol{x}$ it leads to a return $R^0(\boldsymbol{x})$, which can be calculated as indicated in Section 3.4.3.2, using (3.10) or (3.11).

Let $\boldsymbol{g}^1(\boldsymbol{x})$ be the policy defined as the $\boldsymbol{u}$ which optimizes

$$E = r(\boldsymbol{x}, \boldsymbol{u}) + R^0(\boldsymbol{f}(\boldsymbol{x}, \boldsymbol{u})). \tag{3.12}$$

We shall show that $\boldsymbol{g}^1(\boldsymbol{x})$ leads to a return $R^1(\boldsymbol{x})$ which is better than $R^0(\boldsymbol{x})$.

Suppose that a minimum is sought, and that $r(\boldsymbol{x}, \boldsymbol{u}) \geq 0$. Consider the sequence $\pi_m$ of nonstationary policies such that

$$\begin{aligned} \boldsymbol{g}(\boldsymbol{x}_n) &\equiv \boldsymbol{g}^1(\boldsymbol{x}_n), \qquad n \leq m, \\ \boldsymbol{g}(\boldsymbol{x}_n) &\equiv \boldsymbol{g}^0(\boldsymbol{x}_n), \qquad n > m. \end{aligned} \tag{3.13}$$

To each $\pi_m$ corresponds a positive return $\mathscr{R}_m(\boldsymbol{x})$ such that

$$\mathscr{R}_m(\boldsymbol{x}) = \sum_0^{\infty} r(\boldsymbol{x}_n, \boldsymbol{g}(\boldsymbol{x}_n)) = \sum_0^{m} r(\boldsymbol{x}_n, \boldsymbol{g}^1(\boldsymbol{x}_n)) + r(\boldsymbol{x}_{m+1}, \boldsymbol{g}^0(\boldsymbol{x}_{m+1})) + R^0[\boldsymbol{f}(\boldsymbol{x}_{m+1}, \boldsymbol{g}^0(\boldsymbol{x}_{m+1}))],$$

$$\mathscr{R}_{m+1}(\boldsymbol{x}) = \sum_0^{\infty} r(\boldsymbol{x}_n, \boldsymbol{g}(\boldsymbol{x}_n)) = \sum_0^{m} r(\boldsymbol{x}_n, \boldsymbol{g}^1(\boldsymbol{x}_n)) + r(\boldsymbol{x}_{m+1}, \boldsymbol{g}^1(\boldsymbol{x}_{m+1})) + R^0[\boldsymbol{f}(\boldsymbol{x}_{m+1}, \boldsymbol{g}^1(\boldsymbol{x}_{m+1}))].$$

By definition of $\boldsymbol{g}^1(\boldsymbol{x})$, it follows from (3.12) that

$$r(\boldsymbol{x}, \boldsymbol{g}^1(\boldsymbol{x})) + R^0[\boldsymbol{f}(\boldsymbol{x}, \boldsymbol{g}^1(\boldsymbol{x}))] \le r(\boldsymbol{x}, \boldsymbol{g}^0(\boldsymbol{x})) + R^0[\boldsymbol{f}(\boldsymbol{x}, \boldsymbol{g}^0(\boldsymbol{x}))]$$

so that

$$0 \le \mathscr{R}_{m+1} \le \mathscr{R}_m \le \cdots \le \mathscr{R}_1 \le \mathscr{R}_0 = R^0. \tag{3.14}$$

As $m \to \infty$, the policy $\pi_m$ tends to the stationary policy $\boldsymbol{g}^1(\boldsymbol{x})$, and thus

$$0 \le R^1(\boldsymbol{x}) \le R^0(\boldsymbol{x}).$$

Thus $R^1(\boldsymbol{x})$ is an improvement on $R^0(\boldsymbol{x})$, and since $R^1(\boldsymbol{x})$ is bounded, $\boldsymbol{g}^1(\boldsymbol{x})$ is admissible.

If the above process is repeated, a sequence of approximations $\boldsymbol{g}^0(\boldsymbol{x})$, $\boldsymbol{g}^1(\boldsymbol{x}), \ldots, \boldsymbol{g}^k(\boldsymbol{x}), \ldots$, is obtained, with the returns

$$0 \le \hat{R}(\boldsymbol{x}) \le \cdots \le R^k(\boldsymbol{x}) \le \cdots \le R^1(\boldsymbol{x}) \le R^0(\boldsymbol{x}). \tag{3.15}$$

Thus this iterative method converges to the optimal solution $\hat{R}(\boldsymbol{x})$, and for a class of hypotheses much broader than in the case of the recurrence relation (3.6).

### *3.4.5 Conclusion*

Two iterative methods have been described. The first is in return space, and its convergence depends on properties of the operator $\mathscr{F}[\ ]$, which are quite restrictive and difficult to establish. The second method is in policy space, and converges if $r \ge 0$ in the case of search for a minimum, and provided the iteration is begun with an admissible policy. The calculations involved in these methods are diagrammed in Fig. 3.1.

From Fig. 3.1 it can be seen that the second method, $(D, E, F)$, involves two nested loops. However, it may be the preferable method, since the nature of its convergence is known, there is the possibility of convenient initialization, and the intermediate results have some practical significance (in this respect, the method can be compared with the gradient method described by Boudarel *et al.* [1, Vol. 2]).

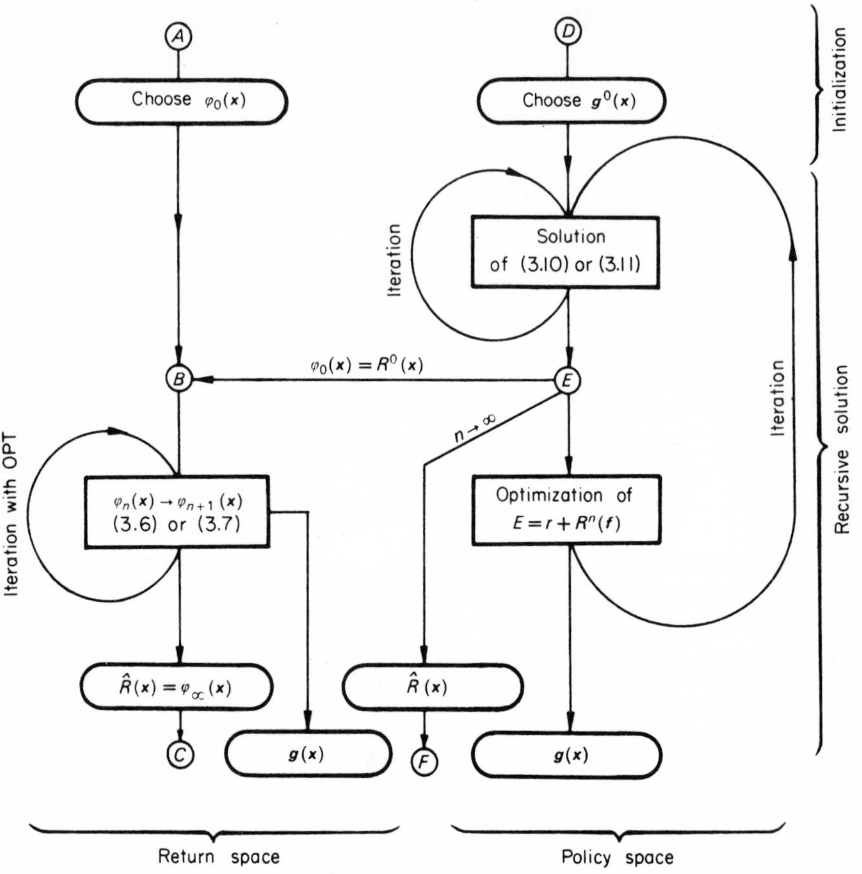


FIG. 3.1. Approximation in return space and in policy space.

# Chapter 4 Practical Solution of the Optimal Recurrence Relation

In the two preceding chapters we have seen that calculation of the optimal control function is related to solution of the optimal recurrence equation

$$\varphi_n(\boldsymbol{x}) = \underset{\boldsymbol{u} \in \Omega}{\text{opt}} \left[ r(\boldsymbol{x}, \boldsymbol{u}) + \varphi_{n-1}(\boldsymbol{f}(\boldsymbol{x}, \boldsymbol{u})) \right]. \tag{4.1}$$

It has been shown that, if $\varphi_{n-1}(\boldsymbol{x})$ is known, calculation of $\varphi_n(\boldsymbol{x})$ for an arbitrary value $\boldsymbol{\alpha}$ of $\boldsymbol{x}$ presents no theoretical difficulties, since the calculation reduces to search for the optimum of the function in brackets in (4.1). However, certain difficulties arise in solution of (4.1) using a digital computer. Principally these involve the search for the optimum in (4.1), storage of the functions $\varphi_n(\boldsymbol{x})$, and the precision of the calculations.

We shall treat the first point only briefly, in Section 4.1. On the other hand, the second difficulty is a major obstacle to application of the method to processes of high order, as discussed in Section 4.2. In order to make full use of the capabilities of computers, and thus to lessen the severity of this obstacle as much as possible, we examine in detail in Section 4.3 the domain in which lie the functions $\varphi_n(\boldsymbol{x})$. This leads to the definition of iterative algorithms in Section 4.4. In Section 4.5 we present a method for economical storage of the functions $\varphi_n(\boldsymbol{x})$. Finally, we consider the case of linear processes: first with quadratic criteria (Section 4.6) leading to a particularly simple literal solution; and then with a terminal criterion which depends only on a limited number of coordinates of $\boldsymbol{x}$ and leads, after a change of variables, to a lowering of the dimensionality of the problem.

## 4.1 Search for an Optimum

As has already been said, the solution of (4.1) requires search for the optimum of

$$\psi(\boldsymbol{u}) = r(\boldsymbol{\alpha}, \boldsymbol{u}) + \varphi_{n-1}(\boldsymbol{f}(\boldsymbol{\alpha}, \boldsymbol{u})), \tag{4.2}$$

for $\boldsymbol{u} \in \Omega$. This is a nonlinear programming problem, as discussed by Boudarel *et al.* [1, Vol. 2, Part 1]. Let us consider here the case of a scalar control.

The most elementary method of optimization is simply to sweep through the allowed interval of variation of $u$, say $[a, b]$, using a step $\delta$, calculating the function (4.1) at each step, and to retain the best value encountered. For the case of maximization, this procedure is indicated in Fig. 4.1. If

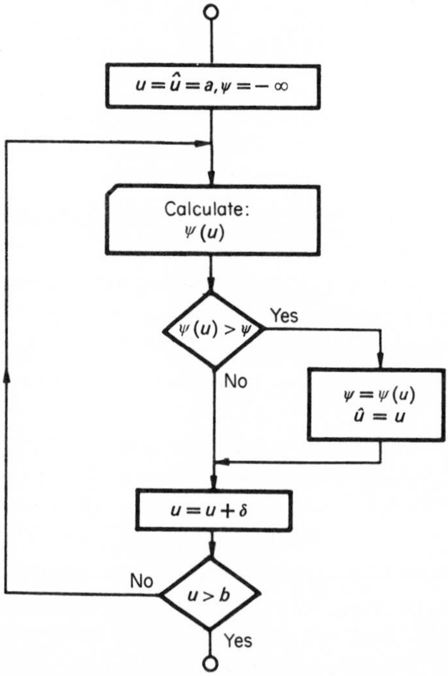


FIG. 4.1. Flow chart for maximum search by simple comparison.

$\psi(u)$ is sufficiently smooth, this procedure locates the optimum with a precision of the order of $\delta/2$, using $(b - a)/\delta$ evaluations of $\psi(u)$.

For the case of a unimodal function, Boudarel *et al.* [1, Vol. 2] presents an algorithm which locates the optimum to within a given precision, using a minimum number of evaluations of $\psi(u)$. This method uses the properties of the Fibonacci number sequence, and the reader is referred to Chapter 5 of the work mentioned for details about the use of this very efficient algorithm.

In the more general case of a vector, or multivariable, control $\boldsymbol{u}$, an iterative method is also used, such as the gradient method or the methods of Newton or Davidon. When $\psi(\boldsymbol{u})$ is either convex or concave, a solution can be obtained quite rapidly. In certain cases the solution is even geometrically evident. It is sometimes also possible to use theoretical results which lead to a literal solution.

In concluding this brief section, we remark that one of the advantages of dynamic programming is that the method reduces to a series of nonlinear programming problems, each relative to a single control $\boldsymbol{u}_n$, and that, as a result, the problem most often remains quite modest.

## 4.2 The Problem of Dimensionality

### 4.2.1 *Principle of a Function Table*

Consider a scalar function $\varphi(\boldsymbol{x})$, with $\boldsymbol{x} \in \mathbf{R}^p$, and with each component of $\boldsymbol{x}$ defined over the domain $0 \leq x_i \leq 1$. If this function is not representable analytically in terms of known functions, one way to conserve the memory of the computer is to tabulate the function values for particular values of $\boldsymbol{x}$. Let $\delta$ be the step length for each coordinate of $\boldsymbol{x}$. Then numerical values of $\varphi(\boldsymbol{x})$ are to be stored in memory for component values of $\boldsymbol{x}$ of the form $x_k = k\delta$, where $k$ is an integer between 0 and $1/\delta = M$. Thus a total of $M^p$ values of $\varphi(\boldsymbol{x})$ must be stored. If it is later necessary to know a value of $\varphi(\boldsymbol{x})$ for an $\boldsymbol{x}$ which is not in the grid of values used, it is necessary to interpolate in the table.

$M$ is taken sufficiently large that interpolation can be carried out with the required precision. In general the functions $\hat{R}(\boldsymbol{x})$ are continuous in $\boldsymbol{x}$, as a consequence of the physical significance of these functions, and this justifies interpolation in a table of $\hat{R}(\boldsymbol{x})$. On the other hand, the control functions $\boldsymbol{g}(\boldsymbol{x}, n)$ are often discontinuous, and as a result, tables obtained by solving (4.1) must be used with care.

### 4.2.2 *The Problem of Precision*

When a function to be tabulated is known in advance, it is possible to determine the step size which assures a desired precision for a given method of interpolation. A variable step size can also be used, which reduces the size of the table. However, the use of (4.1) only allows construction of the table point by point, and since $\varphi_n(\boldsymbol{x})$ is not known in advance, it is difficult to choose the largest step size compatible with a given error tolerance.

It is, however, possible in theory to calculate an upper bound for the error committed in the iterative solution of (4.1) using tables. If no interpolation is done, $\boldsymbol{x}$ being replaced by the value $\boldsymbol{x}^*$ corresponding to the nearest grid point, the formulas have already been established [22]. In the scalar case, if the functions $f_n(\ )$ and $r_n(\ )$ satisfy Lipschitz conditions,

$$|f_n(x', u) - f_n(x'', u)| \leq K_n |x' - x''| a,$$

$$|r_n(x', u) - r_n(x'', u)| \leq K_n' |x' - x''| b,$$

where $x'$ and $x''$ are in the domain $D_n$ of possible values, the upper bound $\varepsilon_n(\delta)$ of the error made in $\varphi_n(x)$ using a grid step $2\delta$ can be found from the following recurrence relations:

$$\varepsilon_n(\delta) = \varepsilon_{n-1}(\delta) + \mu_{n-1}(\delta),$$

$$\mu_{n-1}(\delta) = \mu_{n-2}(K'_{n-1}\, \delta^b) + K_{n-2}\, K'^a_{n-1}\, \delta^{ab}$$

using the initial values

$$\varepsilon_1(\delta) = 0,$$

$$\mu_1(\delta) = K_0\, K_1'^a \delta^{ab}.$$

### 4.2.3 *The Problem of Dimensionality*

We have seen that $M^p$ values of $\varphi(\boldsymbol{x})$ must be stored in memory. This number increases exponentially with $p$. Since these values must be referred to very often in the solution of (4.1), they should be stored in the rapid-access portion of the computer memory. However, the amount of fast memory in a computer is often limited, for the sake of economy, to $2^{15}$ or $2^{16}$ words, i.e., about 32,000 or 64,000 words. For a memory of $2^{15}$ words, entirely filled by the values of $\varphi(\boldsymbol{x})$, Table 4.1 shows the relation between the dimension of $\boldsymbol{x}$ and the allowed grid size used for the table of function values.

For $p$ greater than about 3, the number of values which can be taken for each coordinate is very small. In addition, increasing the memory size from $2^{12}$ to $2^{16}$ only allows $M$ to be increased by $2^{1/p}$, which for $p = 4$ is an increase of only about 20%. Thus the storage problem limits the number of dimensions possible for $\boldsymbol{x}$ rather severely. This limit, encountered in the solution of (4.1), also arises in synthesizing the control function $\boldsymbol{g}(\boldsymbol{x})$.

**Table 4.1**

| Number of dimensions $= p$ | $M = 1/\delta$ |
|---|---|
| 1 | 32,768 |
| 2 | 181 |
| 3 | 32 |
| 4 | 13 |
| 5 | 8 |

Dimensionality is thus an obstacle to the use of dynamic programming with processes of even moderately high order. However, it can be circumvented, at least partially, in certain specific cases, and in general overcome if certain precautions are taken. These precautions are essentially of two types. The domain of definition of $\varphi(\boldsymbol{x})$ should be taken as small as possible for each specific problem. If the intervals are small, only a limited number of points can be used, while maintaining the desired precision. Second, the function $\varphi(\boldsymbol{x})$ should be represented as efficiently as possible, using the table. This topic will be treated using polynomial series representations.

## 4.3 Domain of Definition of the Return Function

In the preceding section the return function was arbitrarily considered on the interval [0, 1]. Actually, however, each variable $x_i$ needs to be considered on some interval $[a_i, b_i]$. Obviously, $a_i$ and $b_i$ must be chosen such that the values of $\boldsymbol{x}$ for which $\varphi(\boldsymbol{x})$ is needed are all on the interval. However, in order to obtain a step $\delta$ as small as possible, since $M$ is fixed by the dimensionality, the interval length $L_i = b_i - a_i$ should be chosen as small as possible. This is a calibration problem.

In some problems the values for $a_i$ and $b_i$ can be chosen *a priori*, considering limits for the physical quantities represented by the variables $x_i$, the components of the state vector $\boldsymbol{x}$. In addition if the control forces are bounded, these limits can also be chosen by considering the reachable region of the state. In problems with the final state free, the reachable regions starting from the possible regions of initial points are considered. On the other hand, in problems with terminal constraints, the regions from which the desired final states can be reached are considered, that is, the controllable regions. These ideas were discussed by Boudarel *et al.* [1, Vol. 1, Chap. 7]. We will apply that material now in the discrete case.

To illustrate the method, let us consider the second-order process

$$\boldsymbol{x}^+ = \begin{bmatrix} 1 & 1 \\ 0 & 1 \end{bmatrix} \boldsymbol{x} + \begin{bmatrix} 1 \\ 2 \end{bmatrix} u, \tag{4.3}$$

where $u$ satisfies the constraint $|u| \leq 1$. To obtain the region $A_n$ reachable at time $n$, starting from the initial region $A$ at time 0 consisting of the single point

$$\boldsymbol{x}_0 = \begin{bmatrix} 0 \\ 3 \end{bmatrix},$$

it is only necessary to apply (4.3) to each point of $A_{n-1}$, using $-1 \leq u \leq 1$. Due to the linearity of (4.3), each point of $A_{n-1}$ generates a line segment $AB$ in $A_n$, as illustrated in Fig. 4.2. Starting from the initial region $A_0 = \{x_0\}$, the regions shown in Fig. 4.3 are obtained in sequence.

To determine the region controllable to $A$ at time $N$, the system (4.3) is used in the reverse direction:

$$\boldsymbol{x}^- = \begin{bmatrix} 1 & 1 \\ 0 & 1 \end{bmatrix}^{-1} \left\{ \boldsymbol{x} - \begin{bmatrix} 1 \\ 2 \end{bmatrix} u \right\} = \begin{bmatrix} 1 & -1 \\ 0 & 1 \end{bmatrix} \boldsymbol{x} + \begin{bmatrix} +1 \\ -2 \end{bmatrix} u. \tag{4.4}$$

This allows the regions $C_{N-u}$ in Fig. 4.4 to be constructed.

If the initial region is some set $A_0$ of initial states, and the final state is to be the origin reached at time $N$, it is necessary to consider both the

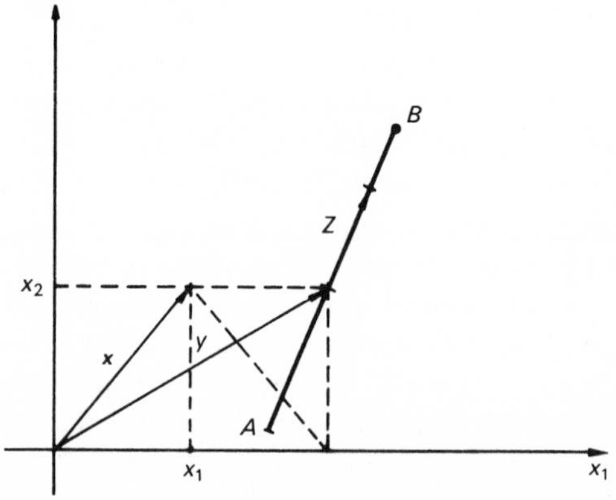


FIG. 4.2. Principle of the construction of the reachable domains.

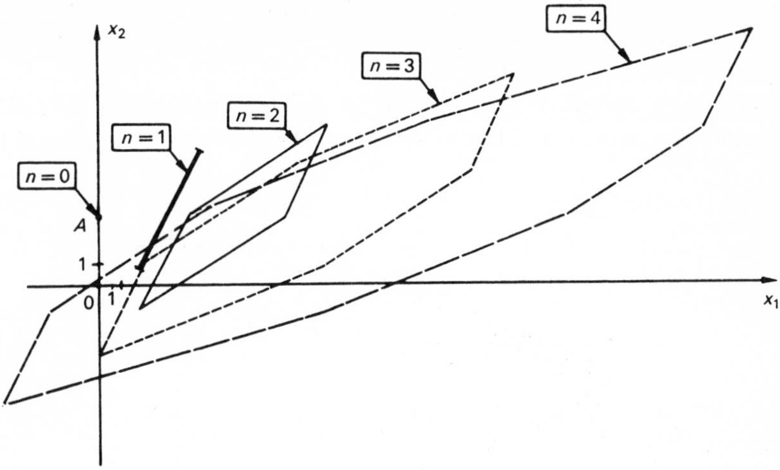


FIG. 4.3. Domains reachable starting from state $A$.

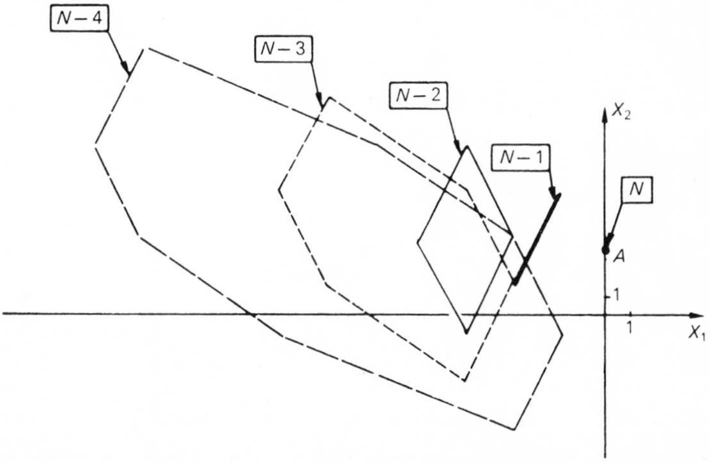


FIG. 4.4. Domains controllable to state $A$.

reachable domains $A_0, \ldots, A_N$, and the controllable regions $C_N, \ldots, C_0$. The admissible trajectories must belong simultaneously to these reachable and controllable domains. That is, at any given time the origin must be reachable from the present state, and the present state must be reachable from the set of given initial states. Thus the set $D_n$ of admissible states

$x_n$ is $D_n = A_n \cap C_n$. For the process (4.3), with $N = 6$, and with $A_0$ a rectangle, the regions $A_n$ are as shown in Fig. 4.5, $C_n$ are as shown in Fig. 4.6, and $D_n$ are as in Fig. 4.7.

It is sufficient to determine $\varphi_n(\boldsymbol{x})$ at points within the regions $D_n$ defined above. Relation (4.1) indeed requires calculation of $\varphi_{n-1}(\boldsymbol{f}(\boldsymbol{x}_n, \boldsymbol{u}_n))$

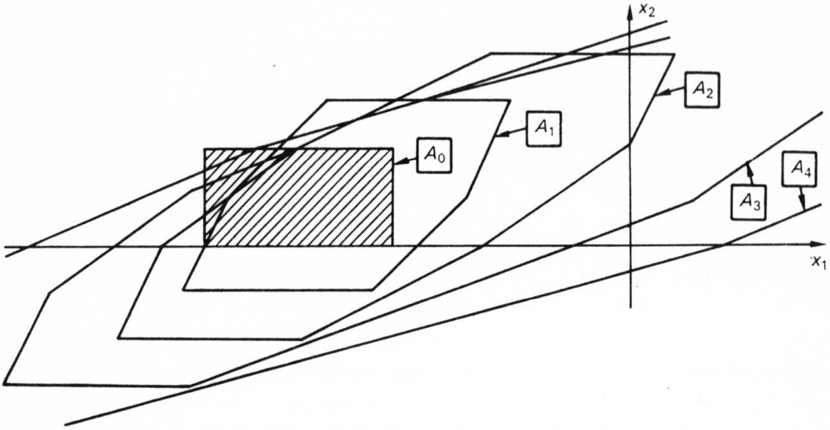


FIG. 4.5. Domains reachable starting from domain $A_0$.

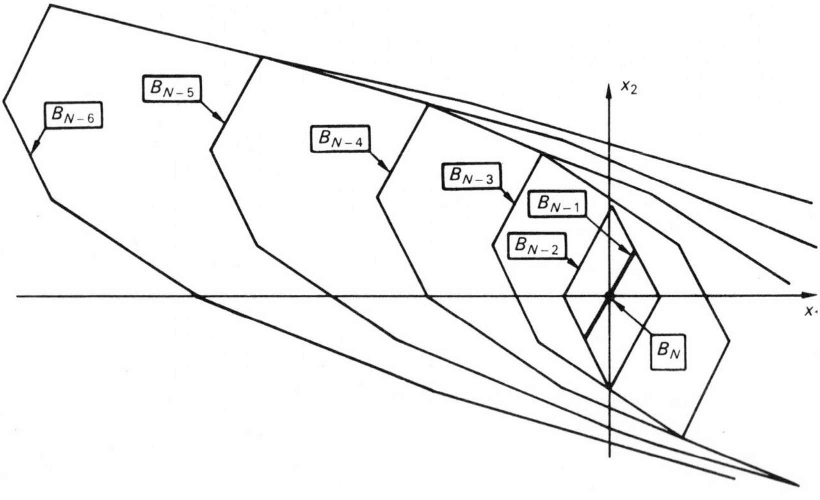


FIG. 4.6. Domains controllable to the origin.

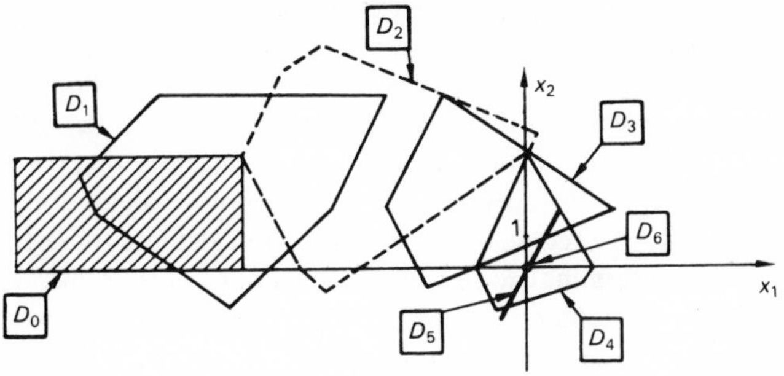


FIG. 4.7. Domains admissible from $A_0$ to the origin in six steps.

for values of $\boldsymbol{f}(\boldsymbol{x}_n, \boldsymbol{u}_n)$ corresponding to the reachable region for $\boldsymbol{x}_n$, which must necessarily be controllable to the final state.

It is difficult to determine the regions $D_n$, and most often the admissible domains are only of theoretical interest.

The optimal trajectories leading from the initial region to the terminal region belong to subsets $\hat{D}_n$ of the regions $D_n$, $\hat{D}_n \subset D_n$. If $\varphi_n(\boldsymbol{x})$ is to be tabulated over as small a region as possible, the region $\hat{D}_n$ should be used. This does not change the solution obtained for (4.1). Preliminary determination of the regions $\hat{D}_n$ requires determination of the optimal trajectories joining the boundaries of the regions of initial and final states. But these are precisely the solutions to the original problem, which we are trying to find using the solutions of (4.1).

## 4.4 Solution by Successive Approximations

As we have just seen, to obtain the most precise solution to the problem, $\varphi_n(\boldsymbol{x})$ should be tabulated over the smallest possible region, $\hat{D}_n$. However, since these latter are a part of the solution to the problem, a method of successive approximations must be used to find them. Two approaches are convenient.

### *4.4.1 The "Wide Grid" Method*

The "wide grid" method involves solution of equation (4.1) using domains for the functions that approach closer and closer to $\hat{D}_n$. The process is started using either the domain of $\boldsymbol{x}_n$, or, in cases in which the reachable and controllable domains can be determined geometrically or numerically, the smaller regions $D_n$.

In general if the number of dimensions is greater than about 4, the table step will be too large (thus the name "wide grid" for the method), and the use of (4.1) leads to a solution which is not sufficiently precise. However, the approximate control law $\tilde{g}(\boldsymbol{x})$ obtained allows calculation of a first approximation to the regions $\hat{D}_n$, using $\tilde{g}(\boldsymbol{x})$ to calculate those trajectories which satisfy the terminal conditions. Then Eq. (4.1) is applied over these smaller regions $\hat{D}_n$, which leads to a better approximation since a smaller step can be used in constructing the table.

If the regions $\hat{D}_n$ are still too large for the number of points allowed in the table, the initial or final regions can be partitioned, and the problem worked separately for each subregion. This allows computing time to be traded for working memory.

### *4.4.2 The Relaxation Method*

The relaxation method is particularly appropriate when a nominal initial state and its immediately adjacent states constitute the initial region. The method begins with selection of a sequence of domains $\Delta_n^0$ sufficiently small to allow use of a convenient table size, these domains being centered on an estimated nominal trajectory.

Additional constraints are then imposed on $\boldsymbol{u}_n$, such that

$$\boldsymbol{x}_{n+1} = \boldsymbol{f}(\boldsymbol{x}_n, \boldsymbol{u}_n, n) \in \Delta_{n+1}^0. \tag{4.5}$$

Application of (4.1) then allows determination of the optimal control law for this modified problem, which necessarily results in trajectories passing through $\Delta_n^0$. If the regions $\Delta_n^0$ chosen happened to be the same as the $\hat{D}_n$, only those trajectories originating on the boundary of $\Delta_0^0$ will pass through the boundary of $\Delta_n^0$. In general, however, this will not be true. A new sequence $\Delta_n^1$ is then constructed by centering each region on the trajectory just found, starting from the nominal initial state. This process is then repeated until the nominal trajectory lies entirely within the regions $\Delta_n^k$.

The domains can then be adjusted such that no trajectory issuing from the region of initial states is deformed by the boundaries of $\Delta_n^k$. This indicates that the additional constraints introduced on the control are inactive, and thus all these trajectories are optimal. Finally, the $\Delta_n^k$ tend to the $\hat{D}_n$.

This procedure is illustrated in Fig. 4.8 for a first-order process.

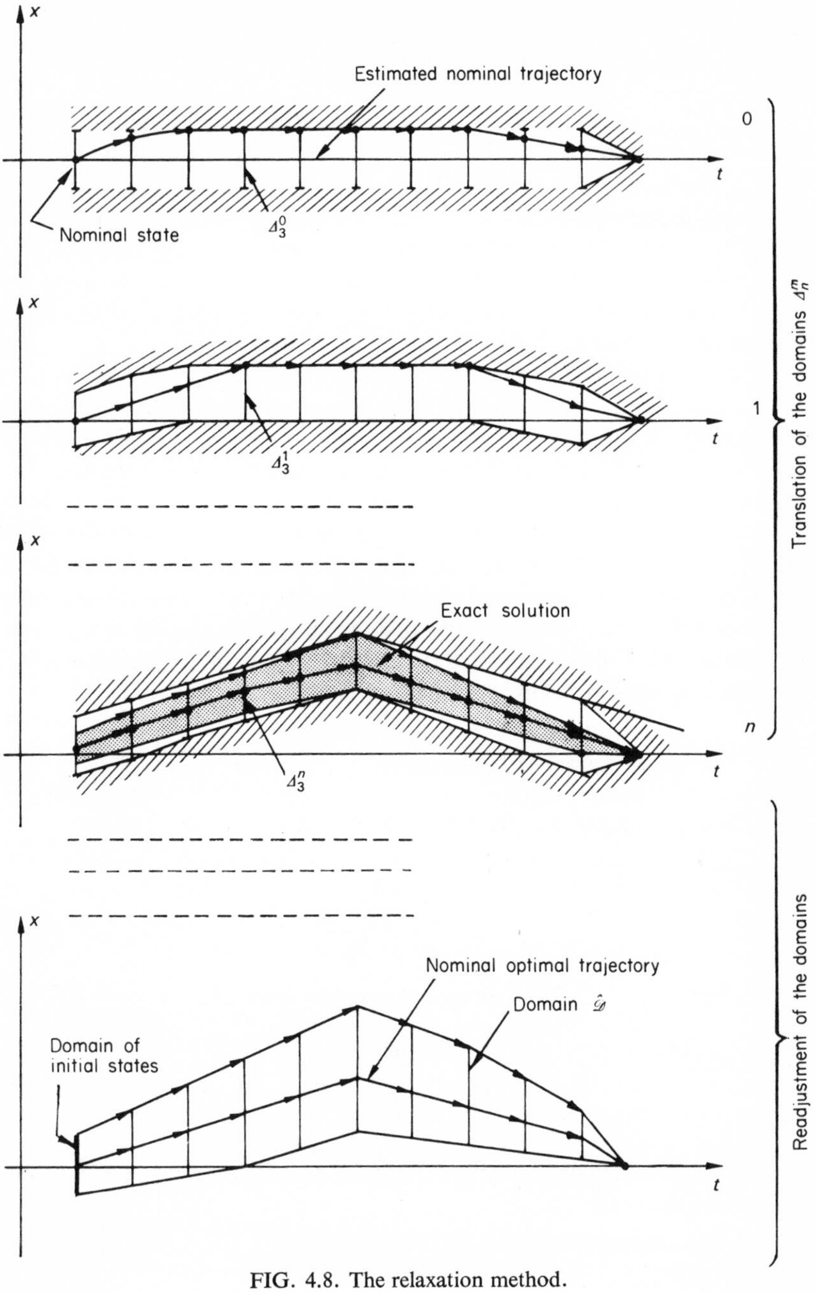


FIG. 4.8. The relaxation method.

## 4.5 The Use of Legendre Polynomials

Tabulation of a function at equally spaced points is certainly not the most economical procedure. For example, a polynomial of degree $d$ can be exactly represented by $d + 1$ numbers, the polynomial coefficients. The following approximate method of representation is based on three main ideas: (1) Approximation of the return functions $\varphi_n(\mathbf{x})$ by polynomials; (2) definition of these polynomials at points chosen so as to facilitate the solution of (4.1); and (3) use of the orthogonality properties of the polynomials in finding the coefficients.

If the functions $\varphi(\mathbf{x})$ are to be approximated over a given interval in such a way that the mean squared error is minimum, an expansion in terms of Legendre polynomials is appropriate. We shall show that it is always possible to replace the relation (4.1) by a recursion relation between the expansion coefficients in the approximation of $\varphi(\mathbf{x})$. In order to establish this result, we shall review briefly the properties of the Legendre polynomials (recursive definition and orthogonality), and an application of these properties to numerical integration. Then we consider the one-dimensional case and establish the relation in question using the properties developed. Finally, the method is extended to the multidimensional case, and the improvement in the amount of storage necessary is indicated.

### *4.5.1 Review of Legendre Polynomials*

The Legendre polynomials $P_n(x)$ satisfy the recurrence relation

$$P_{n+1}(x) = \frac{2n+1}{n+1}\, xP_n(x) - \frac{n}{n+1}\, P_{n-1}(x), \qquad P_0(x) = 1, \qquad P_1(x) = x. \tag{4.6}$$

They can be shown to be orthogonal over the interval $[-1, 1]$:

$$\int_{-1}^{1} P_m(x)P_l(x)\,\mathrm{d}x = \begin{cases} \dfrac{2}{2m+1}, & l = m \\ 0, & l \neq m. \end{cases} \tag{4.7}$$

Any continuous square-integrable function $f(x)$ can be decomposed into a series of the form

$$f(x) = \sum_{i=0}^{\infty} a^i P_i(x), \tag{4.8}$$

where the $a^i$, because of the orthogonality property, are given by

$$a^i = \frac{2i+1}{2} \int_{-1}^{1} f(x) P_i(x)\, dx. \tag{4.9}$$

The $P_i(x)$ form a complete orthogonal basis for the space of functions square-integrable on the interval $[-1, 1]$. The truncated sum

$$f_n(x) = \sum_{0}^{n} a^i P_i(x) \tag{4.10}$$

is the best least-squares approximation of $f(x)$ by a polynomial of degree $n$ over the interval $[-1, 1]$.

The Legendre polynomials have been applied to the evaluation of integrals in a way which we shall use for the calculation of (4.9). With the following formula, an integral can be replaced by a sum, in which only a set of values of the function to be integrated, taken at particular points, appears:

$$\int_{-1}^{1} f(x)\, dx = \sum_{i=1}^{r} H_i f(x_i) + \varepsilon. \tag{4.11}$$

Here $x_i$ is the $i$th root of $P_r(x)$:

$$P_r(x_i) = 0.$$

The constants $H_i$ are given by

$$H_i = \frac{2(1 - x_i^2)}{(r+1)^2 [P_{r+1}(x_i)]^2}. \tag{4.12}$$

The error $\varepsilon$ is zero if $f(x)$ is a polynomial of degree at most $2r - 1$, otherwise it is of order $2r$ in $x$. If $\varepsilon$ is neglected, (4.11) is an approximate formula.

### 4.5.2 *Application to First-Order Processes*

With a first-order process we wish to solve

$$\varphi_{n+1}(x) = \max_{u \in \Omega} [r(x, u) + \varphi_n(f(x, u))], \qquad u, x \in \mathbf{R}. \tag{4.13}$$

Suppose that $x$ is defined on the interval $[-1, 1]$, and let us approximate the function $\varphi_n(x)$ by its expansion in Legendre polynomials of degree up to $m$:

$$\tilde{\varphi}_n(x) = \sum_{j=0}^{m} a_n{}^j P_j(x). \tag{4.14}$$

Suppose now that the $a_n{}^j$ in (4.14) are known. A recursive procedure, yielding the $a_{n+1}^j$ from the $a_n{}^j$, is possible, using the formulas of the preceding paragraph. From (4.9) and (4.10),

$$a_{n+1}^j = \frac{2j+1}{2} \int_{-1}^{1} \tilde{\varphi}_{n+1}(x) P_j(x)\, dx = \frac{2j+1}{2} \sum_{i=1}^{r} H_i \tilde{\varphi}_{n+1}(x_i) P_j(x_i), \tag{4.15}$$

with no error in the case that

$$2r - 1 \geq m + j \rightarrow r \geq m + 1 .$$

In addition, $\tilde{\varphi}_{n+1}(x)$ is to satisfy (4.13):

$$\tilde{\varphi}_{n+1}(x_i) = \max_{u \in \Omega} [r(x_i, u) + \sum_{j=0}^{m} a_n{}^j P_j(f(x_i, u))] . \tag{4.16}$$

All the quantities $x_i$, $H_i$, $P_j(x_i)$ can be calculated in advance. The polynomial values

$$P_i(f(x_i, u)) = V_i$$

can be calculated recursively using (4.6):

$$V_{i+1} = \frac{2i+1}{i+1} V_1 V_i - \frac{i}{i+1} V_{i-1} , \qquad V_0 = 1, \qquad V_1 = f(x_i, u) . \tag{4.17}$$

The use of (4.17) with different values of $u$, along with (4.16), allows calculation of the $\tilde{\varphi}_{n+1}(x_i)$, from the $a_n{}^j$, and from these, using (4.15), the $a_{n+1}^j$ follow. These calculations are summarized in the flow chart of Fig. 4.9. If the interval of definition is different from $[-1, 1]$, a simple change of variable reduces the actual problem to one on that interval.

Calculation of the $\tilde{\varphi}_n(x_i)$ leads to the $u_i$ relative to the $x_i$. Thus the function $g(x)$ is found at $r$ points, and thus can also be approximated by a polynomial of degree $m$:

$$\tilde{g}(x) = \sum_{j=0}^{m} b_j P_j(x), \tag{4.18}$$

with

$$b_j = \frac{2j+1}{2} \sum_{i=1}^{r} H_i \hat{u}_i P_j(x_i) . \tag{4.19}$$

**Remark.** Although $R(x)$ is almost always continuous, the same is not the case for $g(x)$. In case $g(x)$ is not continuous, the use of Legendre polynomials may lead to an approximation which is too poor, particularly near the points of discontinuity.

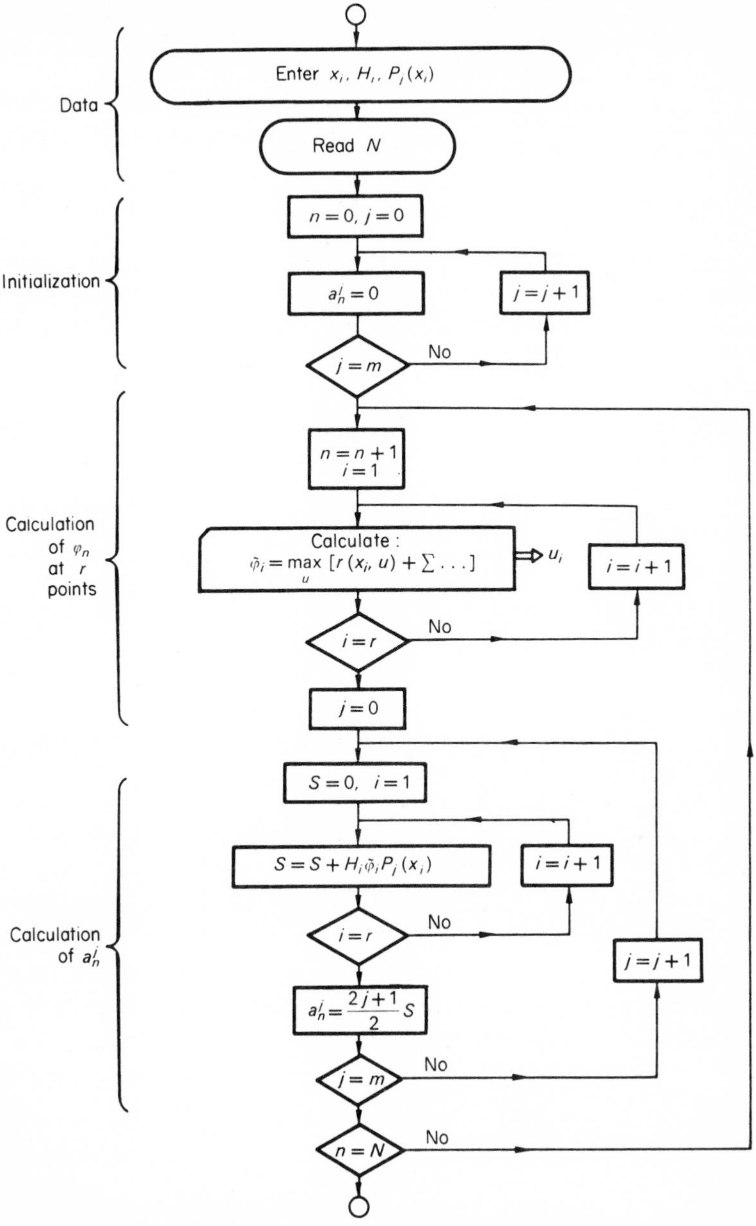


FIG. 4.9. Flow chart for the calculation of optimal return functions using Legendre polynomials.

### 4.5.3 *The General Case*

The above method can be extended to the case in which $\boldsymbol{x}$ is multidimensional. In that case, $\varphi(\boldsymbol{x})$ is approximated by a series of the form

$$\tilde{\varphi}_n(x_1, \ldots, x_l) = \sum_{j_1=0}^{m} \cdots \sum_{j_l=0}^{m} a_n^{j_1 \cdots j_l} P_{j_1}(x_1) \cdots P_{j_l}(x_l). \tag{4.20}$$

The coefficients $a^{j_1 \cdots j_l}$ are again calculated recursively using formulas analogous to (4.15) and (4.16). For each $n$ the number of coefficients is $(m+1)^l$. Table 4.2 indicates the amount of storage required to hold these

**Table 4.2**

| $l$ | Legendre polynomials | | | | | | Grid with 100 points per coordinate |
|---|---|---|---|---|---|---|---|
| | $m = 3$ | $m = 4$ | $m = 5$ | $m = 6$ | $m = 8$ | $m = 10$ | |
| 1 | 4 | 5 | 6 | 7 | 9 | 11 | 100 |
| 2 | 16 | 25 | 36 | 49 | 81 | 121 | 10,000 |
| 3 | 64 | 125 | 216 | 313 | 719 | 1,331 | |
| 4 | 256 | 625 | 1,296 | 2,401 | 6,561 | 14,641 | |
| 5 | 1,024 | 3,125 | 7,776 | 16,807 | | | |
| 6 | 4,096 | 15,625 | | | | | |
| 7 | 16,384 | | | | | | |

coefficients for various $l$ and $m$. It can be seen that even with a limit of 32,000 words of memory, with $m$ small, processes of reasonable order can be treated. For example, with $m = 3$, a process with $l = 6$, corresponding to motion of a weight in space, requires only 4096 coefficients.

### 4.5.4 *Synthesis of the Controller*

When the control function itself is also expanded in a series of Legendre polynomials, it can be computed in real time by again making use of the relation (4.6). The appropriate flow chart is shown in Fig. 4.10. The

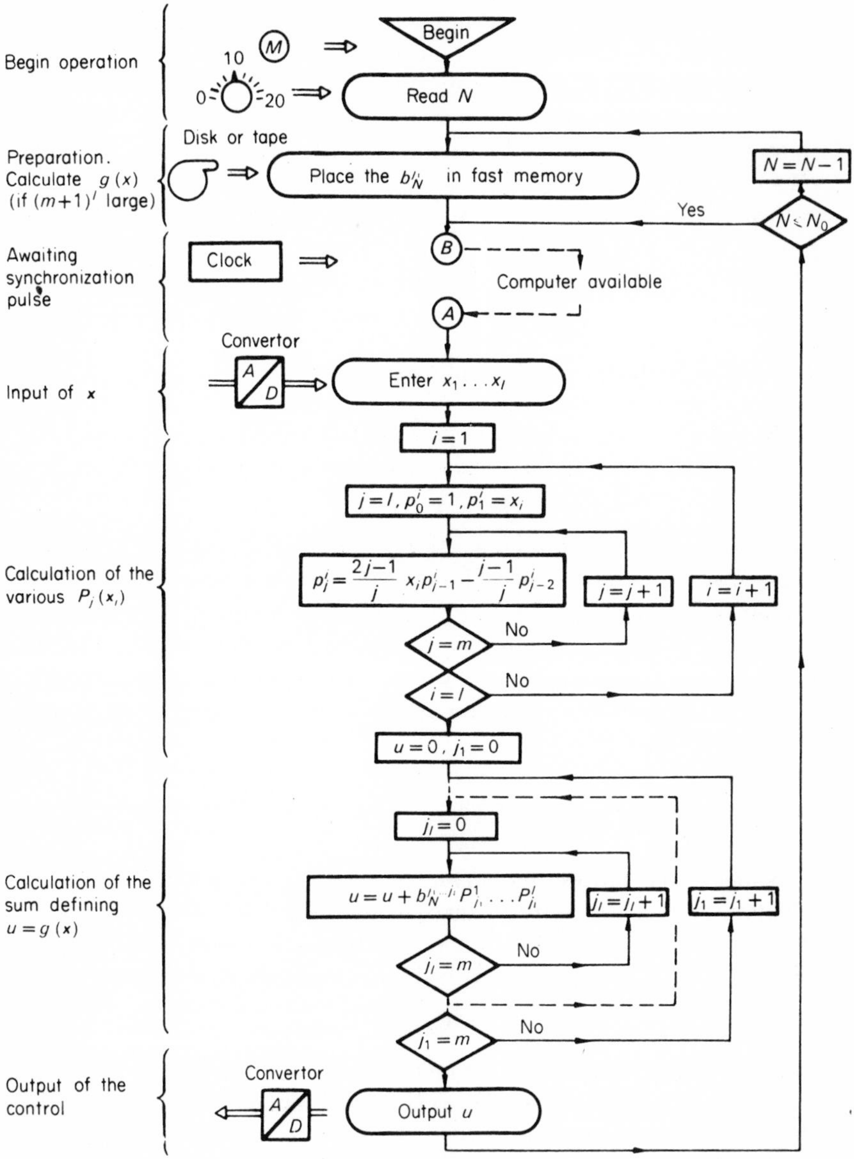


FIG. 4.10. Flow chart of a controller based on an expansion in Legendre polynomials.

coefficients $b_n^{j_1 \cdots j_l}$ of $g_n(\boldsymbol{x})$ are calculated off-line using formula (4.19), and stored in a peripheral bulk memory. Only the coefficients relative to the current value of $n$ are transferred into the working memory.

Upon arrival of a priority interrupt signal from the synchronization clock, computation begins at point $A$ on the flow chart. The value of the state $\boldsymbol{x}$ is first entered, and then the corresponding polynomial values $P_j(x_i)$ are calculated using the recursion (4.6). The sum analogous to (4.20) is then found, which gives $u = g_n(\boldsymbol{x})$, which can then be applied to the process. After loading the next set of coefficients, $b_{n+1}^{j_1 \cdots j_l}$, the computer is available for other uses until arrival of the next clock pulse.

## 4.6 Linear Systems with Quadratic Costs

In linear processes where $\boldsymbol{u}$ is unconstrained and the cost functions are quadratic forms in $\boldsymbol{x}$ and $\boldsymbol{u}$, the optimal returns are quadratic forms in $\boldsymbol{x}$:

$$\hat{R}(\boldsymbol{x}, n) = \boldsymbol{x}^{\mathrm{T}} \mathbf{Q}_n \boldsymbol{x} .$$

Recurrence relations exist among the $\mathbf{Q}_n$. In this case, only $l(l+1)/2$ words of memory are needed for their storage, since the $\mathbf{Q}_n$ are symmetric, and the storage requirements are thus greatly reduced. In addition, the control law is linear in $\boldsymbol{x}$, which facilitates its storage, and its calculation by the controller. Thus, in this special case, which is the problem of classical control theory, a simple solution can be obtained. We shall now develop these results.

### *4.6.1 Definitions and Notation*

Suppose the process terminates at time $N$. The optimal recurrence equation evolves backward in time, so that it is convenient to introduce the time variable $m$, related to real time $n$ by $n = N - m$. With this convention, a linear process is described by an equation of the form

$$\boldsymbol{x}_{m-1} = \mathbf{F}_m \boldsymbol{x}_m + \mathbf{H}_m \boldsymbol{u}_m . \tag{4.21}$$

For conciseness of notation, the index $m$ will be dropped, and $\mathbf{F}_m$ written as $\mathbf{F}$, $\boldsymbol{x}_{m-1}$ as $\boldsymbol{x}^-$, $\boldsymbol{u}_{m+1}$ as $\boldsymbol{u}^+$, $\mathbf{H}_m$ as $\mathbf{H}$, etc. This is not meant to imply that the system is stationary but is solely a matter of notational convenience.

The cost criterion being quadratic, the return corresponding to the initial state $\boldsymbol{x}$ at the time $N - m$, i.e., $m$ steps before the end of the process, can be written

$$\hat{R}(\boldsymbol{x}, N - m) = \hat{R}_m(\boldsymbol{x}) = \sum_{i=m}^{1} [\boldsymbol{x}_i{}^{\mathrm{T}}\mathbf{A}_i\boldsymbol{x}_i + 2\boldsymbol{x}_i{}^{\mathrm{T}}\mathbf{B}_i\boldsymbol{u}_i + \boldsymbol{u}_i{}^{\mathrm{T}}\mathbf{C}_i\boldsymbol{u}_i], \tag{4.22}$$

where $\mathbf{A}_i$ and $\mathbf{C}_i$ are symmetric matrices.

**Remark.** For some problems, the elementary return $r_i$ depends on $\boldsymbol{u}_i$ and $\boldsymbol{x}_{i-1}$, so that

$$R_m(\boldsymbol{x}) = \sum_{i=m}^{1} [\boldsymbol{x}_{i-1}^{\mathrm{T}}\mathbf{A}_i'\boldsymbol{x}_{i-1} + 2\boldsymbol{x}_{i-1}^{\mathrm{T}}\mathbf{B}_i'\boldsymbol{u}_i + \boldsymbol{u}_i{}^{\mathrm{T}}\mathbf{C}_i'\boldsymbol{u}_i]. \tag{4.23}$$

This case can be put in the form (4.22) by using

$$\begin{aligned}\mathbf{A} &= \mathbf{F}^{\mathrm{T}}\mathbf{A}'\mathbf{F},\\ \mathbf{B} &= \mathbf{F}^{\mathrm{T}}[\mathbf{A}'\mathbf{H} + \mathbf{B}'],\\ \mathbf{C} &= \mathbf{H}^{\mathrm{T}}\mathbf{A}'\mathbf{H} + \mathbf{H}^{\mathrm{T}}\mathbf{B}' + \mathbf{B}'\mathbf{H}^{\mathrm{T}} + \mathbf{C}'.\end{aligned}$$

### *4.6.2 The Optimal Recurrence Equation*

Using (4.21) and (4.22) in (4.1) leads to

$$\hat{R}(\boldsymbol{x}) = \underset{\boldsymbol{u}}{\mathrm{opt}}[\boldsymbol{x}^{\mathrm{T}}\mathbf{A}\boldsymbol{x} + 2\boldsymbol{x}^{\mathrm{T}}\mathbf{B}\boldsymbol{u} + \boldsymbol{u}^{\mathrm{T}}\mathbf{C}\boldsymbol{u} + \hat{R}^{-}(\mathbf{F}\boldsymbol{x} + \mathbf{H}\boldsymbol{u})]. \tag{4.24}$$

If we now assume that $\hat{R}^{-}(\boldsymbol{x})$ is a quadratic form,

$$\hat{R}^{-}(\boldsymbol{x}) = \boldsymbol{x}^{\mathrm{T}}\mathbf{Q}^{-}\boldsymbol{x},$$

we have

$$\hat{R}^{-}(\mathbf{F}\boldsymbol{x} + \mathbf{H}\boldsymbol{u}) = (\boldsymbol{x}^{\mathrm{T}}\mathbf{F}^{\mathrm{T}} + \boldsymbol{u}^{\mathrm{T}}\mathbf{H}^{\mathrm{T}})\mathbf{Q}^{-}(\mathbf{F}\boldsymbol{x} + \mathbf{H}\boldsymbol{u}).$$

Since $\boldsymbol{u}$ is unconstrained, if the problem has a solution it must be such that the partial derivative of the expression in brackets in (4.24) with respect to $\boldsymbol{u}$ vanishes:

$$2\mathbf{B}^{\mathrm{T}}\boldsymbol{x} + 2\mathbf{C}\hat{\boldsymbol{u}} + 2\mathbf{H}^{\mathrm{T}}\mathbf{Q}\mathbf{F}\boldsymbol{x} + 2\mathbf{H}^{\mathrm{T}}\mathbf{Q}^{-}\mathbf{H}\hat{\boldsymbol{u}} = 0.$$

This yields

$$\hat{\boldsymbol{u}} = -[\mathbf{C} + \mathbf{H}^{\mathrm{T}}\mathbf{Q}^{-}\mathbf{H}]^{-1}[\mathbf{B}^{\mathrm{T}} + \mathbf{H}^{\mathrm{T}}\mathbf{Q}^{-}\mathbf{F}]\boldsymbol{x}, \tag{4.25}$$

so that the optimal control law is linear,

$$\hat{\boldsymbol{u}} = -\mathbf{L}\boldsymbol{x}, \tag{4.26}$$

where the (nonconstant) matrix $\mathbf{L}$ is

$$\mathbf{L} = [\mathbf{C} + \mathbf{H}^{\mathrm{T}}\mathbf{Q}^{-}\mathbf{H}]^{-1}[\mathbf{B}^{\mathrm{T}} + \mathbf{H}^{\mathrm{T}}\mathbf{Q}^{-}\mathbf{F}]. \tag{4.27}$$

Substituting $\hat{\boldsymbol{u}}$ into the bracketed expression in (4.24), it will be found that $\hat{R}(\boldsymbol{x})$ is also a quadratic form, with matrix

$$\begin{aligned}\mathbf{Q} &= \mathbf{A} - 2\mathbf{BL} + \mathbf{L}^{\mathrm{T}}\mathbf{CL} + \mathbf{F}^{\mathrm{T}}\mathbf{Q}^{-}\mathbf{F} - 2\mathbf{F}^{\mathrm{T}}\mathbf{Q}^{-}\mathbf{HL} + \mathbf{L}^{\mathrm{T}}\mathbf{H}^{\mathrm{T}}\mathbf{Q}^{-}\mathbf{HL} \\ &= \mathbf{A} + \mathbf{F}^{\mathrm{T}}\mathbf{Q}^{-}\mathbf{F} - [\mathbf{B} + \mathbf{F}^{\mathrm{T}}\mathbf{Q}^{-}\mathbf{H}]\mathbf{L},\end{aligned}$$

which is to say

$$\mathbf{Q} = \mathbf{A} + \mathbf{F}^{\mathrm{T}}\mathbf{Q}^{-}\mathbf{F} - [\mathbf{B} + \mathbf{F}^{\mathrm{T}}\mathbf{Q}^{-}\mathbf{H}][\mathbf{C} + \mathbf{H}^{\mathrm{T}}\mathbf{Q}^{-}\mathbf{H}]^{-1}[\mathbf{B}^{\mathrm{T}} + \mathbf{H}^{\mathrm{T}}\mathbf{Q}^{-}\mathbf{F}]. \tag{4.28}$$

### 4.6.3 *Initialization of the Recurrence Equation*

We have shown that if $\hat{R}_{m-1}(\boldsymbol{x})$ is a quadratic form, so also is $\hat{R}_m(\boldsymbol{x})$, and that the corresponding matrix satisfies the recurrence relation (4.28). It only remains to show that the return $\hat{R}_k(\boldsymbol{x})$ with which the iteration should be started is itself also a quadratic form.

If the final state of the system is free, it is clear that

$$\hat{R}_0(\boldsymbol{x}) \equiv 0, \tag{4.29}$$

and (4.17) is initialized with $\mathbf{Q}_0 = 0$, the null matrix. If the process is to terminate at a specified state,

$$\boldsymbol{x}_0 = \boldsymbol{x}_{\mathrm{f}},$$

the final $k$ commands must satisfy

$$\boldsymbol{x}_{\mathrm{f}} = \mathbf{F}_1\mathbf{F}_2 \cdots \mathbf{F}_k\boldsymbol{x}_k + \mathbf{H}_1\boldsymbol{u}_1 + \mathbf{F}_1\mathbf{H}_2\boldsymbol{u}_2 + \cdots + \mathbf{F}_1\mathbf{F}_2 \cdots \mathbf{F}_{k-1}\mathbf{H}_k\boldsymbol{u}_k. \tag{4.30}$$

This is a linear system for the $\boldsymbol{u}_i$. Depending on $\mathbf{F}_i$, $\mathbf{H}_i$, and $k$, there can be no solution, a unique solution, or infinitely many solutions. Which of these cases occurs is determined by the rank of the matrix

$$\mathbf{M}_k = [\mathbf{H}_1, \mathbf{F}_1\mathbf{H}_2, \mathbf{F}_1\mathbf{F}_2\mathbf{H}_3, \ldots, \mathbf{F}_1 \cdots \mathbf{F}_{k-1}\mathbf{H}_k]. \tag{4.31}$$

If $\boldsymbol{u}$ has $q$ components, and $\boldsymbol{x}$ has $p$ components, then

$$\operatorname{rank}[\mathbf{M}_k] < kq \le p \Rightarrow \text{no solution} \tag{4.32}$$

$$p = \operatorname{rank}[\mathbf{M}_k] = kq \Rightarrow \text{unique solution} \tag{4.33}$$

$$p = \operatorname{rank}[\mathbf{M}_k] < kq \Rightarrow \text{many solutions}. \tag{4.34}$$

Boudarel *et al.* [1, Vol. 1, Sect. 4.2.1], discuss the concept of controllability. Controllability of the process to the desired final state corresponds precisely to the question considered here of the existence of a control sequence $\boldsymbol{u}_1, \ldots, \boldsymbol{u}_k$. In particular, in the stationary case it is necessary and sufficient that $\mathbf{M}_p$ be of rank $p$. If this condition is satisfied, there then exists a $k \leq p$ such that either (4.33) or (4.34) holds. To determine the return in terms of $\boldsymbol{u}_1, \ldots, \boldsymbol{u}_k$, it is necessary to treat these two cases separately.

### 4.6.3.1 *Unique Solution*

There is a unique solution if $u$ is scalar and the process is controllable. In the multidimensional case, it is necessary that $p$ be a multiple of $q$ $(qk = p)$, and that $|\mathbf{M}_k| \neq 0$. This is a rather special case, however, which occurs when the system is normal, for example.

In this case,

$$\begin{bmatrix} \boldsymbol{u}_1 \\ \vdots \\ \boldsymbol{u}_k \end{bmatrix} = \mathbf{M}_k^{-1}(\boldsymbol{x}_\mathrm{f} - \mathbf{F}_1 \cdots \mathbf{F}_k \boldsymbol{x}_k). \tag{4.35}$$

The $\boldsymbol{u}_i$ are thus linear in $\boldsymbol{x}_\mathrm{f}$ and $\boldsymbol{x}_k$, as well as in the $\boldsymbol{x}_i$.

Substituting these controls into the cost criterion, there results, for a given $\boldsymbol{x}_0$:

$$\hat{R}_k(\boldsymbol{x}) = \alpha_k + 2\boldsymbol{\beta}_k{}^\mathrm{T}\boldsymbol{x} + \boldsymbol{x}^\mathrm{T}\mathbf{Q}_k\boldsymbol{x}. \tag{4.36}$$

For this to be a quadratic form, it is necessary that $\boldsymbol{x}_0$ be the origin. If the final state $\boldsymbol{x}_\mathrm{f}$ is an equilibrium point, it can be taken as the origin, and the problem becomes one in which the final state is the origin. In this case the cost $\hat{R}_k(\boldsymbol{x})$ for $k = p/q$ is a quadratic form as claimed.

### 4.6.3.2 *Multiple Solutions*

Let us consider the smallest $k \leq p$ for which (4.34) holds. From among the various sequences $\boldsymbol{u}_1, \ldots, \boldsymbol{u}_k$ which will carry the system to the desired final state, we wish to select that leading to the minimum return. This problem was already treated by Boudarel *et al.* [1, Vol. 2], using nonlinear programming. They established (Problem 2 of Part 4) that the optimal control law was again linear in $\boldsymbol{x}_0$ and $\boldsymbol{x}_k$. Thus the multiple solutions case is analogous to the unique solution case, so far as the structure of the optimal return is concerned.

In Chapter 8 we shall return to the details of the calculation of $\mathbf{Q}_k$ and of the regulator sequence $\boldsymbol{u}_1, \ldots, \boldsymbol{u}_k$. We recall, however, that for a stationary process it has been established (Boudarel *et al.* [1, Vol. 1, Sect. 6.6.1]), that the regulator sequence is stationary and of the form

$$\boldsymbol{u}_m = \mathbf{L}\boldsymbol{x}_m, \tag{4.37}$$

where $\mathbf{L}$ is precisely the linear law, relating $\boldsymbol{u}_k$ and $\boldsymbol{x}_k$, deduced from (4.35):

$$(4.35) \Rightarrow \boldsymbol{u}_k = \mathbf{L}\boldsymbol{x}_k \Rightarrow \boldsymbol{u}_m = \mathbf{L}\boldsymbol{x}_m, \qquad 1 \le m \le k.$$

### 4.6.4 Final State Not an Equilibrium Point

If the final state is not an equilibrium point, the optimal return is no longer quadratic but has the complete form (4.36). It is then necessary to correct the calculations of Section 4.6.2, using

$$\hat{R}^-(\boldsymbol{x}) = \alpha^- + 2\boldsymbol{\beta}^{-\mathrm{T}}\boldsymbol{x} + \boldsymbol{x}^{\mathrm{T}}\mathbf{Q}^-\boldsymbol{x}.$$

It can easily be shown that for this case

$$\hat{\boldsymbol{u}} = -\mathbf{L}\boldsymbol{x} - \boldsymbol{\lambda}, \tag{4.38}$$

with $\mathbf{L}$ given by (4.27), and with

$$\boldsymbol{\lambda} = [\mathbf{C} + \mathbf{H}^{\mathrm{T}}\mathbf{Q}^-\mathbf{H}]^{-1}\mathbf{H}^{\mathrm{T}}\boldsymbol{\beta}^-. \tag{4.39}$$

As before, $\mathbf{Q}$ satisfies the recurrence relation (4.28), while

$$\boldsymbol{\beta} = (\mathbf{F} - \mathbf{HL})^{\mathrm{T}}\boldsymbol{\beta}^-, \tag{4.40}$$

$$\alpha = \alpha^- - \boldsymbol{\beta}^{-\mathrm{T}}\mathbf{H}[\mathbf{C} + \mathbf{H}^{\mathrm{T}}\mathbf{Q}^-\mathbf{H}]^{-1}\mathbf{H}^{\mathrm{T}}\boldsymbol{\beta}^-, \tag{4.41}$$

$$\alpha = \alpha^- - \boldsymbol{\lambda}^{\mathrm{T}}\mathbf{H}^{\mathrm{T}}\boldsymbol{\beta}^-.$$

Thus the term of the optimal control law which is linear in $\boldsymbol{x}$ is the same as when the final state is an equilibrium point. Superimposed on the linear part there is an open-loop control $\boldsymbol{\lambda}$ which depends on $\boldsymbol{\beta}$. This latter satisfies (4.40), which is the adjoint system corresponding to the closed-loop process using the control (4.26).

### *4.6.5 Conclusion*

We have seen that the problem considered in this section has a particularly simple solution. It is only necessary

1. To find the minimum $k$ for which the final state can be reached from an arbitrary state. This first step determines the final portion of the control law, and the form of the corresponding cost.
2. To use the matrix recursion relations (4.28) and (4.26) to determine the control law in terms of the matrices $\mathbf{L}_n$. If the final state is not an equilibrium point, (4.39) and (4.40) must be used to calculate the additive term $\boldsymbol{\lambda}$.

As stated at the beginning of this section, the optimal return function and the corresponding control law can both be represented in terms of matrices. It is then possible to treat high-order problems without being limited by memory volume required.

Step 1 can be eliminated by replacing the terminal constraint by a quadratic penalty function adjoined to the cost. This is the case of a free final state, with

$$R_0(\boldsymbol{x}) = K(\boldsymbol{x} - \boldsymbol{x}_f)^T(\boldsymbol{x} - \boldsymbol{x}_f), \tag{4.42}$$

which is to say

$$\alpha_0 = K\boldsymbol{x}_f{}^T\boldsymbol{x}_f, \qquad \boldsymbol{\beta}_0 = -K\boldsymbol{x}_f, \qquad \mathbf{Q}_0 = K\mathbf{1}. \tag{4.43}$$

The use of this artifice can lead to difficulties in the choice of $K$. At best, the terminal constraint is only approximately satisfied, while at worst, the evaluation of (4.28) may be in error.

**Remark.** In calculating $\boldsymbol{u}$ we have taken no account of second order conditions. In fact, it is necessary that $\mathbf{H}^T\mathbf{Q}^-\mathbf{H}$ be nonnegative definite. The same must thus be true for $\mathbf{Q}^-$, which will always be the case regardless of the control strategy if the elementary return is positive, in the search for a minimum. This requires that

$$\begin{bmatrix} \mathbf{A} & \mathbf{B} \\ \mathbf{B}^T & \mathbf{C} \end{bmatrix}$$

be nonnegative definite.

## 4.7 Linearization

In certain problems, rocket guidance for example, the collection of initial states of interest is localized around some nominal state. By methods other than dynamic programming, discussed by Boudarel *et al.* [1, Vols. 2 and 4], the optimal control law $\boldsymbol{u}_n$ and the corresponding trajectory $\boldsymbol{x}_n$ can be found relative to the nominal initial state. If the actual optimal trajectory begins at an initial state near the nominal, the equations of motion of the system can be linearized around the nominal trajectory.

Thus, letting

$$\boldsymbol{\varepsilon}_n = \boldsymbol{x}_n - \mathring{\boldsymbol{x}}_n,$$

$$\boldsymbol{v}_n = \boldsymbol{u}_n - \mathring{\boldsymbol{u}}_n,$$

there results

$$\boldsymbol{\varepsilon}_{n+1} = \mathbf{F}_x(\mathring{\boldsymbol{x}}_n, \mathring{\boldsymbol{u}}_n, n)\boldsymbol{\varepsilon}_n + \mathbf{F}_u(\mathring{\boldsymbol{x}}_n, \mathring{\boldsymbol{u}}_n, n)\boldsymbol{v}_n. \tag{4.44}$$

In the same way, the elementary return can be expanded including the quadratic term.

The problem thus reduces to that of a linear system with quadratic cost, for which the methods of the preceding paragraphs lead to a simple numerical solution. Hence, linearization around a nominal trajectory allows a linear numerical control system to be used.

## 4.8 Reduction of the Dimensionality

In certain problems it is possible to reduce the dimensionality by means of an appropriate change of variables. A particularly important case is that in which the process is linear, and the cost is made up of a terminal cost which depends on only $l$ coordinates of $\boldsymbol{x}$ (which can always be assumed to be the first $l$), and an elementary return which depends only on the control $\boldsymbol{u}_n$. With the numbering conventions used above, these conditions can be written

$$\boldsymbol{x}_{m-1} = \mathbf{F}\boldsymbol{x}_m + \mathbf{H}\boldsymbol{u}_m,$$

$$R_k = \sum_{m=k}^{1} r_m(\boldsymbol{u}_m) + \rho(x_0^{\,1}, \ldots, x_0^{\,l}). \tag{4.45}$$

Let us write $\boldsymbol{x}_0$ as a function of the initial state $\boldsymbol{x}_k$ and the control $\boldsymbol{u}_n$:

$$\boldsymbol{x}_0 = \boldsymbol{\Phi}_k \boldsymbol{x}_k + \sum_{m=k}^{1} \boldsymbol{\Psi}_m \boldsymbol{u}_m, \tag{4.46}$$

where

$$\boldsymbol{\Phi}_k = \mathbf{F}_1 \cdots \mathbf{F}_k \rightarrow \boldsymbol{\Phi}_k = \boldsymbol{\Phi}_{k-1} \mathbf{F}_k, \tag{4.47}$$

$$\boldsymbol{\Psi}_k = \mathbf{F}_1 \cdots \mathbf{F}_{k-1} \mathbf{H}_k \rightarrow \boldsymbol{\Psi}_k = \boldsymbol{\Phi}_{k-1} \mathbf{H}_k. \tag{4.48}$$

Let $\boldsymbol{x}_0'$ denote the vector having for components the first $l$ elements of $\boldsymbol{x}_0$. According to (4.46), $\boldsymbol{x}_0'$ is linear in the $\boldsymbol{u}_m$:

$$\boldsymbol{x}_0' = \boldsymbol{b} + \sum_{m=k}^{1} \mathbf{K}_m \boldsymbol{u}_m, \tag{4.49}$$

where

$$\boldsymbol{b} = l \text{ first elements of } \boldsymbol{\Phi}_k \boldsymbol{x}_k, \tag{4.50}$$

$$\mathbf{K}_m = l \text{ first rows of } \boldsymbol{\Psi}_m. \tag{4.51}$$

The cost can then be written

$$R_k = \sum_{m=k}^{1} r_m(\boldsymbol{u}_m) + \rho(\boldsymbol{x}_0'). \tag{4.52}$$

The relation (4.49) can be written recursively as

$$\boldsymbol{x}_{m-1}' = \boldsymbol{x}_m' + \mathbf{K}_m \boldsymbol{u}_m, \tag{4.53}$$

$$\boldsymbol{x}_k' = \boldsymbol{b}. \tag{4.54}$$

The initial problem (4.45) can then be replaced by the equivalent problem (4.52), (4.53), involving a state vector $\boldsymbol{x}'$ having only $l$ components.

Solution of this problem by dynamic programming involves introduction of a return $\hat{R}_k(\boldsymbol{x}')$ which is a function of the $l$ components of $\boldsymbol{x}'$, and which satisfies the optimal recurrence equation

$$\hat{R}_k(\boldsymbol{x}') = \underset{\boldsymbol{u} \in \Omega_k}{\operatorname{opt}} [r_k(\boldsymbol{u}) + \hat{R}_{k-1}(\boldsymbol{x}' + \mathbf{K}_k \boldsymbol{u})]. \tag{4.55}$$

This is initialized with

$$\hat{R}_0(\boldsymbol{x}') = \rho(\boldsymbol{x}'), \tag{4.56}$$

and leads to the control law

$$\boldsymbol{u}_k = \boldsymbol{g}_k(\boldsymbol{x}_k') = \boldsymbol{g}_k(\boldsymbol{\Phi}_k' \boldsymbol{x}_k). \tag{4.57}$$

In this $\mathbf{\Phi}_k'$ is the matrix consisting of the first $l$ rows of $\mathbf{\Phi}_k$, and the second equality results from (4.54) and (4.50).

Thus solution of the problem (4.45) requires (1) calculation of $\mathbf{\Phi}_k'$, using the recursion corresponding to (4.47):

$$\mathbf{\Phi}_k' = \mathbf{\Phi}_{k-1}' \mathbf{F}_k, \qquad \mathbf{\Phi}_0' = \mathbf{0}; \tag{4.58}$$

(2) calculation of $\mathbf{K}_k$, using the relation deduced from (4.48):

$$\mathbf{K}_k = \mathbf{\Phi}_{k-1}' \mathbf{H}_k; \tag{4.59}$$

(3) solution of the optimal recurrence equation (4.55), which leads to the optimal control functions $\boldsymbol{g}_k(\ )$.

**Remark 1.** If $\rho(\ )$ depends only on one coordinate, $\mathbf{\Phi}_k'$ reduces to a row vector $\boldsymbol{\phi}_k^{\mathrm{T}}$, which satisfies the recursion

$$\boldsymbol{\varphi}_k^{\mathrm{T}} = \boldsymbol{\varphi}_{k-1}^{\mathrm{T}} \mathbf{F}_k. \tag{4.60}$$

The adjoint system can be recognized here.

**Remark 2.** If the $\boldsymbol{u}_k$ are unconstrained and the cost is quadratic,

$$R_k = \sum_{m=k}^{1} \boldsymbol{u}_m^{\mathrm{T}} \mathbf{C}_m \boldsymbol{u}_m + \boldsymbol{x}_0'^{\mathrm{T}} \mathbf{S} \boldsymbol{x}_0', \tag{4.61}$$

the recursion (4.55) can be solved using the results of Section 4.6. In this case

$$\hat{R}_k(\boldsymbol{x}') = \boldsymbol{x}'^{\mathrm{T}} \mathbf{Q}_k \boldsymbol{x}', \qquad \mathbf{Q}_0 = \mathbf{S}. \tag{4.62}$$

Formulas (4.58), (4.59) and (4.27), (4.28) lead to the system

$$\begin{aligned} \mathbf{\Phi}_k' &= \mathbf{\Phi}_{k-1}' \mathbf{F}_k, \qquad \mathbf{\Phi}_0' = \mathbf{0}, \\ \mathbf{K}_k &= \mathbf{\Phi}_k' \mathbf{H}_k, \\ \mathbf{L}_k &= [\mathbf{C}_k + \mathbf{K}_k^{\mathrm{T}} \mathbf{Q}_{k-1} \mathbf{K}_k]^{-1} \mathbf{K}_k \mathbf{Q}_{k-1}, \\ \mathbf{Q}_k &= \mathbf{Q}_{k-1}(\mathbf{1} - \mathbf{K}_k^{\mathrm{T}} \mathbf{L}_k), \qquad \mathbf{Q}_0 = \mathbf{S}. \end{aligned}$$

The optimal control law is

$$\hat{\boldsymbol{u}}_k = \mathbf{K}_k \mathbf{\Phi}_k' \boldsymbol{x}_k.$$

**Remark 3.** This method may also be used if $l$ coordinates of $\boldsymbol{x}_0$ are specified as a terminal constraint, rather than appearing in a terminal cost. Further, in all problems in which terminal constraints require that $\boldsymbol{x}$ reach some subspace of $l$ dimensions, a change of variables will reduce the problem to the form considered above.

# PART 2 Discrete Random Processes

# Chapter 5 General Theory

## 5.1 Optimal Control of Stochastic Processes

### *5.1.1 Stochastic Processes*

The idea of a stochastic process was introduced by Boudarel *et al.* [1, Vol. 1, Chap. 10], starting from a discrete deterministic process

$$x_{n+1} = f(x_n, u_n, n), \tag{5.1}$$

with an additive total return, and having an elementary return

$$r_n = r(x_n, u_n, n). \tag{5.2}$$

The most direct approach is to adjoin to (5.1) and (5.2) terms $e_n$ which are random variables. Then $x_{n+1}$ and $r_n$ become random variables, the statistical parameters of which depend on $x_n$, $u_n$, and the statistics of $e_n$.

An alternate approach is to consider the future state and the elementary return directly as random variables, governed by a conditional probability distribution

$$P(y, a; x_n, \mathbf{u}_n, n) = \text{prob}\{x_{n+1} \leq y; \quad r \leq a\} \qquad \text{if at instant } n$$

$$\text{state is } x_n \text{ and applied control is } u_n. \tag{5.3}$$

This is the transition probability of a discrete Markov process with control. Its properties have been studied by Boudarel *et al.* [1, Vol. 1, Ch. 13].

The use of the function (5.3) does not imply that $x_n$ itself need be perfectly known at time $n$. In the important case in which it is not entirely measurable, by which is meant that the state can not be completely determined by measurements, $\boldsymbol{x}_n$ will be a random variable. We shall return to this point at the end of the chapter.

A larger body of mathematics exists related to the second approach, and it lends itself better to theoretical developments. The second approach will thus be used exclusively in the remainder of this chapter. To use the second approach, however, it is necessary to pass through the first approach as an intermediate step. The calculation of (5.3) from (5.1), (5.2), and the probability distribution of the $\boldsymbol{e}_n$ generally encounters great numerical difficulties. Only in the case of a linear process with Gaussian perturbations having a rational spectrum does a simple solution exist (see Boudarel *et al.* [1, Vol. 1, Chap. 13, Sect. 3.1]).

### 5.1.2 *Definition of an Optimal Control*

Starting from any particular instant $n$, the future evolution of the process up to time $N$ is random. The same is thus true for the total future return,

$$R_n^N = \sum_{k=n}^{N-1} r_k . \tag{5.4}$$

This quantity can thus no longer be optimized, and no longer can serve as a criterion. The average, or expected return, however, is a definite number which can be optimized over the control policy. It will thus be chosen as the cost criterion:

$$\text{criterion} = \mathrm{E}[\text{future return}], \tag{5.5}$$

where the symbol E indicates mathematical expectation.

Two types of optimal control policies can then be defined:

a. Open-loop control: For a known initial state $\boldsymbol{x}_0$, *a priori* control laws

$$\boldsymbol{u}_0 = \tilde{\boldsymbol{g}}_0(\boldsymbol{x}_0), \ldots, \boldsymbol{u}_n = \tilde{\boldsymbol{g}}_n(\boldsymbol{x}_0), \ldots, \boldsymbol{u}_{N-1} = \tilde{\boldsymbol{g}}_{N-1}(\boldsymbol{x}_0), \tag{5.6}$$

are sought, such that the mathematical expectation of the total return between times 0 and $N$ is optimum.

b. Closed-loop control: The state $\boldsymbol{x}_n$ at time $n$ is assumed to be perfectly known. Only the control $\boldsymbol{u}_n$ to be applied at that time is then determined, taking account of the information about $\boldsymbol{x}_n$, in the form

$$\boldsymbol{u}_n = \hat{\boldsymbol{g}}_n(\boldsymbol{x}_n) . \tag{5.7}$$

The optimal closed-loop policy is then specified by the sequence of functions

$$\boldsymbol{u}_0 = \hat{\boldsymbol{g}}_0(\boldsymbol{x}_0), \ldots, \boldsymbol{u}_n = \hat{\boldsymbol{g}}_n(\boldsymbol{x}_n), \ldots, \boldsymbol{u}_{N-1} = \hat{\boldsymbol{g}}_{N-1}(\boldsymbol{x}_{N-1}) . \tag{5.8}$$

Policy b is analogous to that already found in the deterministic case, and diagrammed in Fig. 2.5. Policy a, which does not take account of the states actually attained, leads "on the average" to an effective return less desirable than that obtained with b. In addition, if state-dependent constraints are present, of the form $\boldsymbol{u}_n \in \Omega(\boldsymbol{x}_n)$, an open-loop control policy can result in control commands which are incompatible with the constraints, and which are thus no longer sensible. Thus, we shall formulate the optimal control problem with an eye toward closed-loop control.

So far as the method of solution is concerned, policy a corresponds to a nonlinear programming problem (Boudarel *et al.* [1, Vol. 2]), and can be converted to a policy of type b by noting that, for $n = 0$,

$$\tilde{\boldsymbol{g}}_0(\boldsymbol{x}_0) \equiv \hat{\boldsymbol{g}}_0(\boldsymbol{x}_0) . \tag{5.9}$$

This identity can be used for arbitrary $m$, if $m$ is considered to be the initial time for a process with horizon $[m, N]$. This approach leads to a separate nonlinear programming problem for each $m$. We shall see that, on the contrary, dynamic programming allows the sought solution to be found directly.

In concluding this introduction, it should be noted that, if $\boldsymbol{x}_m$ is not entirely measurable, so that the information obtained at time $m$ does not completely determine the state, it is possible in calculating the closed-loop control $\boldsymbol{u}_m$ to take account of all measurements made between times 0 and $m$. This complicates the structure of the controller and the calculations necessary to specify it. This important point will be taken up at the end of the chapter, and ultimately examined in detail for the case of a linear system.

## 5.2 Processes with Bounded Horizon and Measurable State

By applying the principle of optimality, we shall establish a recurrence relation for the optimal return functions. We shall then discuss the general case of this relation, and finally consider some particular problems.

### 5.2.1 The Basic Equation

Let us consider a closed-loop control law. If it is optimal, the mathematical expectation of the return $R_n^N$ will be a function which depends only on $\boldsymbol{x}_n$ and $n$. We shall denote this function by $\varphi_n(\boldsymbol{x}_n)$.

The return $R_n^N$ can be decomposed as

$$R_n^N = r_n + R_{n+1}^N . \tag{5.10}$$

Its mathematical expectation will thus be the sum of the mathematical expectation of the current return, which is a random variable depending only on $\boldsymbol{x}_n$ and $\boldsymbol{u}_n$, and that of the return from time $n+1$ until time $N$. For a given $\boldsymbol{x}_{n+1}$, if the optimal control policy is used, this latter average return will be just the function $\varphi_{n+1}(\boldsymbol{x}_{n+1})$. However, at time $n$, $\boldsymbol{x}_{n+1}$ is a random variable which depends on $\boldsymbol{u}_n$ and $\boldsymbol{x}_n$.

Thus the introduction of $\varphi_{n+1}(\boldsymbol{x}_{n+1})$ reduces the calculation of the mathematical expectation of $R_n^N$ to that of $r_n + \varphi_{n+1}(\boldsymbol{x}_{n+1})$, which for a given $\boldsymbol{x}_n$ depends only on $\boldsymbol{u}_n$, and relates only to the transition $n \to n+1$.

The principle of optimality now states that $R_n^N$ will be optimal, and thus equal to $\varphi_n(\boldsymbol{x}_n)$, if $\boldsymbol{u}_n$ optimizes the above quantity. Thus

$$\varphi_n(\boldsymbol{x}_n) = \text{opt}\, \mathrm{E}[r_n + \varphi_{n+1}(\boldsymbol{x}_{n+1})] . \tag{5.11}$$

This recurrence relation differs formally from that relative to the deterministic case only in the presence of the expectation operator. It is to be initialized with

$$\varphi_N(\boldsymbol{x}_n) = 0 , \tag{5.12}$$

which corresponds to a free final state, the only case possible for a stochastic process.

In the following sections we shall eliminate the expectation operator, which leads us back to a numerical problem analogous to that treated in Chapter 4, and allows the laws $\hat{\boldsymbol{g}}_n(\boldsymbol{x}_n)$ to be calculated.

### 5.2.2 The General Case

Let us consider the process corresponding to the transition probability (5.3). Let $\mathscr{E}$ be the domain of possible values of the pair $(\boldsymbol{y}, a)$.

Recall first that if the random variable $\boldsymbol{x}$, defined over a domain $\mathscr{D}$, has the probability distribution

$$P(z) = \text{prob}\{\boldsymbol{x} \le \boldsymbol{z}\} , \tag{5.13}$$

then the mathematical expectation of a function $\gamma(x)$ is

$$\mathrm{E}[\gamma(z)] = \int_{\mathscr{D}} \gamma(z)\,\mathrm{d}P(z). \tag{5.14}$$

If (5.14) is taken as a Stieltjes integral, a unique value results, whatever may be the continuity properties of $P(z)$.

Applying this result to (5.11), and taking account of (5.3), yields

$$\varphi_n(\boldsymbol{x}) = \underset{\boldsymbol{u} \in \Omega_n}{\mathrm{opt}} \int_{\mathscr{E}} [a + \varphi_{n+1}(\boldsymbol{y})]\,\mathrm{d}P(\boldsymbol{y}, a; \boldsymbol{x}, \boldsymbol{u}, n). \tag{5.15}$$

This integral, once having been evaluated, determines a function $\psi_n(\boldsymbol{x}, \boldsymbol{u})$, depending only on $\boldsymbol{x}$ and $\boldsymbol{u}$. Thus

$$\varphi_n(\boldsymbol{x}) = \underset{\boldsymbol{u} \in \Omega_n}{\mathrm{opt}}\, \psi_n(\boldsymbol{x}, \boldsymbol{u}). \tag{5.16}$$

For each value $\boldsymbol{x} = \boldsymbol{\alpha}$, this last relation leads to the optimal value of $\mathbf{u}$, which then allows $\varphi_n(\ )$ and $\hat{\boldsymbol{g}}_n(\ )$ to be calculated point by point. As in the deterministic case, the optimum to be found is relative to only a single $\boldsymbol{u}_n$ at each step, and thus only an elementary nonlinear programming problem is involved.

### *5.2.3 A Special Case*

Consider a process which can evolve through any one of $k$ deterministic laws, the choice of the law to be used at each time $n$ being random, and independent of the preceding choices. For such a system,

$$\begin{aligned} \boldsymbol{x}_{n+1} &= \boldsymbol{f}^i(\boldsymbol{x}_n, \boldsymbol{u}_n, n), \\ r_n &= r^i(\boldsymbol{x}_n, \boldsymbol{u}_n, n), \end{aligned} \tag{5.17}$$

where a particular $i$ is chosen with probability $p_i$. Then

$$\mathrm{E}[r_n] = \sum_{i=1}^{k} p_i\, r^i(\boldsymbol{x}_n, \boldsymbol{u}_n, n),$$

$$\mathrm{E}[\varphi_{n+1}(\boldsymbol{x}_{n+1})] = \sum_{i=1}^{k} p_i\, \varphi_{n+1}(\boldsymbol{f}^i(\boldsymbol{x}_n, \boldsymbol{u}_n, n)),$$

and the recurrence relation (5.11) becomes

$$\varphi_n(\boldsymbol{x}) = \underset{\boldsymbol{u} \in \Omega_n}{\mathrm{opt}} \sum_{i=1}^{k} p_i[r^i(\boldsymbol{x}, \boldsymbol{u}, n) + \varphi_{n+1}(\boldsymbol{f}^i(\boldsymbol{x}, \boldsymbol{u}, n))]. \tag{5.18}$$

This equation differs but little from that of the deterministic case. Only the amount of calculation is different, being greater since $\varphi_{n+1}(\,)$ must be evaluated $k$ times, rather than once.

### 5.2.4 *Application to Linear Systems with Quadratic Cost*

Consider the preceding case, but suppose that

$$f^i(\boldsymbol{x}_n, \boldsymbol{u}_n, n) = \mathbf{F}_i \boldsymbol{x} + \mathbf{H}_i \boldsymbol{u}, \tag{5.19}$$

$$r^i(\boldsymbol{x}_n, \boldsymbol{u}_n, n) = \boldsymbol{x}^{\mathrm{T}} \mathbf{A}_i \boldsymbol{x} + 2\boldsymbol{x}^{\mathrm{T}} \mathbf{B}_i \boldsymbol{u} + \boldsymbol{u}^{\mathrm{T}} \mathbf{C}_i \boldsymbol{u}, \tag{5.20}$$

where for brevity the index $n$ is suppressed. As in the deterministic case, assume that the optimal return is a quadratic form in the state $\boldsymbol{x}$:

$$\varphi_n(\boldsymbol{x}) = \boldsymbol{x}^{\mathrm{T}} \mathbf{Q}_n \boldsymbol{x}.$$

It then follows that

$$\begin{aligned}
\boldsymbol{x}^{\mathrm{T}}\mathbf{Q}\boldsymbol{x} &= \operatorname*{opt}_{\boldsymbol{u}}\Big\{\boldsymbol{x}^{\mathrm{T}}\Big(\sum_{i=1}^{k} p_i \mathbf{A}_i\Big)\boldsymbol{x} + 2\boldsymbol{x}^{\mathrm{T}}\Big(\sum_{i=1}^{k} p_i \mathbf{B}_i\Big)\boldsymbol{u} + \boldsymbol{u}^{\mathrm{T}}\Big(\sum_{i=1}^{k} p_i \mathbf{C}_i\Big)\boldsymbol{u} \\
&\qquad + \sum_{i=1}^{k} p_i(\mathbf{F}_i\boldsymbol{x} + \mathbf{H}_i\boldsymbol{u})^{\mathrm{T}}\mathbf{Q}^{+}(\mathbf{F}_i\boldsymbol{x} + \mathbf{H}_i\boldsymbol{u})\Big\} \\
&= \operatorname*{opt}_{\boldsymbol{u}}\Big\{\boldsymbol{x}^{\mathrm{T}}\Big[\sum_{i=1}^{k} p_i(\mathbf{A}_i + \mathbf{F}_i^{\mathrm{T}}\mathbf{Q}^{+}\mathbf{F}_i)\Big]\boldsymbol{x} + 2\boldsymbol{x}^{\mathrm{T}}\Big[\sum_{i=1}^{k} p_i(\mathbf{B}_i + \mathbf{F}_i^{\mathrm{T}}\mathbf{Q}^{+}\mathbf{H}_i)\Big]\boldsymbol{u} \\
&\qquad + \boldsymbol{u}^{\mathrm{T}}\Big[\sum_{i=1}^{k} p_i(\mathbf{C}_i + \mathbf{H}_i^{\mathrm{T}}\mathbf{Q}^{+}\mathbf{H}_i)\Big]\boldsymbol{u}\Big\}.
\end{aligned} \tag{5.21}$$

Thus we have obtained an expression analogous to that of Section 4.6. If $\varphi_{n+1}(\boldsymbol{x})$ is a quadratic form, so also is $\varphi_n(\boldsymbol{x})$, and since $\varphi_N(\boldsymbol{x})$ is identically zero, our hypothesis is justified. The optimal control law, obtained by differentiating the above equation with respect to $\boldsymbol{u}$, is linear:

$$\hat{\boldsymbol{u}} = -\mathbf{L}\boldsymbol{x}, \tag{5.22}$$

where

$$\mathbf{L} = \Big[\sum_i p_i(\mathbf{C}_i + \mathbf{H}_i^{\mathrm{T}}\mathbf{Q}^{+}\mathbf{H}_i)\Big]^{-1}\Big[\sum_i p_i(\mathbf{B}_i^{\mathrm{T}} + \mathbf{H}_i^{\mathrm{T}}\mathbf{Q}^{+}\mathbf{F}_i)\Big]. \tag{5.23}$$

Substituting this control back into the optimal recurrence equation (5.21), a recurrence relation results for $\mathbf{Q}$, initialized with

$$\mathbf{Q}_N = \mathbf{0}, \tag{5.24}$$

which is

$$\begin{aligned}\mathbf{Q} = {} & \Big[\sum_i p_i \mathbf{A}_i\Big] + \Big[\sum_i p_i \mathbf{F}_i^{\mathrm{T}}\mathbf{Q}^+\mathbf{F}_i\Big] \\ & - \Big[\sum_i p_i(\mathbf{B}_i + \mathbf{F}_i^{\mathrm{T}}\mathbf{Q}^+\mathbf{H}_i)\Big]\Big[\sum_i p_i(\mathbf{C}_i + \mathbf{H}_i^{\mathrm{T}}\mathbf{Q}^+\mathbf{H}_i)\Big]^{-1} \\ & \times \Big[\sum_i p_i(\mathbf{B}_i^{\mathrm{T}} + \mathbf{H}_i^{\mathrm{T}}\mathbf{Q}^+\mathbf{F}_i)\Big].\end{aligned} \tag{5.25}$$

Thus, in the case of many possible laws for the system evolution, the structure of the optimal controller and the relation giving the corresponding return are analogous to those of the deterministic case.

### *5.2.5 Stochastic Control Systems*

In defining the problem it was implicitly assumed that the closed-loop control policy, as in the deterministic case, consisted in associating with each value of the state the corresponding optimal control. In the theory of games, however, it is customary to use mixed strategies in which the decisions are chosen randomly, the distribution function governing this action of fate having been chosen optimally depending on the situation.

The transition to this control philosophy can be made by defining a stochastic control policy, by means of a distribution function $\tilde{\omega}(\boldsymbol{u}; \boldsymbol{x}, n)$, the probability of choosing a control less than $\boldsymbol{u}$, if at time $n$ the state is $\boldsymbol{x}$. The optimization is then carried out over the functions $\tilde{\omega}$, which can be defined on some domains $\Omega_n(\boldsymbol{x}_n)$. We shall now show that this additional complication is to no purpose here, since the deterministic law already found, using (5.15), is again the optimum.

Application of the principle of optimality yields

$$\varphi_n'(\boldsymbol{x}) = \operatorname*{opt}_{\tilde{\omega} \in \Pi} \int_{\Omega_n} \left\{ \int_{\mathscr{E}} [a + \varphi_{n+1}'(\boldsymbol{y})] \, \mathrm{d}P(\boldsymbol{y}, a; \boldsymbol{x}, \boldsymbol{u}, n) \right\} \mathrm{d}\tilde{\omega}(\boldsymbol{u}; \boldsymbol{x}, n) \tag{5.26}$$

where $\Pi$ is the set of distribution functions defined on $\Omega_n$, and $\varphi_n'(\boldsymbol{x})$ is the average return obtained using the optimal stochastic policy.

Let us now show that $\varphi_n'(\boldsymbol{x}) \equiv \varphi_n(\boldsymbol{x})$, this being true for $n = N$. The term in braces in (5.26) is the function $\psi_n(\boldsymbol{x}, \boldsymbol{u})$, and (5.26) can thus be written

$$\varphi_n'(\boldsymbol{x}) = \underset{\tilde{\omega} \in \Pi}{\operatorname{opt}} \int_{\Omega_n} \psi_n(\boldsymbol{x}, \boldsymbol{u})\, \mathrm{d}\tilde{\omega}(\boldsymbol{u}; \boldsymbol{x}, n). \tag{5.27}$$

Let $\hat{\boldsymbol{u}}$ be the value of $\boldsymbol{u}$ which optimizes $\psi_n(\boldsymbol{x}, \boldsymbol{u})$ on the domain $\Omega_n$. In the case of search for a maximum, whatever may be the function $\tilde{\omega}$, the inequality

$$\int_{\Omega_n} \psi_n(\boldsymbol{x}, \boldsymbol{u})\, \mathrm{d}\tilde{\omega}(\boldsymbol{u}; \boldsymbol{x}, n) \leq \psi_n(\boldsymbol{x}, \hat{\boldsymbol{u}}) \int_{\Omega_n} \mathrm{d}\tilde{\omega}(\boldsymbol{u}; \boldsymbol{x}, n) = \psi_n(\boldsymbol{x}, \hat{\boldsymbol{u}})$$

holds, since, by definition,

$$\int_{\Omega_n} \mathrm{d}\tilde{\omega}(\boldsymbol{u}; \boldsymbol{x}, n) = 1.$$

Thus

$$\varphi_n'(\boldsymbol{x}) \leq \underset{\boldsymbol{u} \in \Omega_n}{\operatorname{opt}} \psi_n(\boldsymbol{x}, \boldsymbol{u}) = \varphi_n(\boldsymbol{x}), \tag{5.28}$$

which shows that a stochastic policy yields at best a return equal to that possible with a deterministic policy, and that equality of the two returns occurs precisely for the case that the function $\tilde{\omega}$ degenerates to a deterministic law.

## 5.3 Processes with Random Horizon and Measurable State

In most stochastic problems, the final time $N$ is not known in advance, but rather depends, for example, on satisfaction of certain conditions on $\boldsymbol{x}$ of the type

$$T_N(\boldsymbol{x}) = 0. \tag{5.29}$$

There is then a certain probability $\pi(\boldsymbol{x}_n, \boldsymbol{u}_n)$ that $\boldsymbol{x}_{n+1}$ will be such that $T(\boldsymbol{x}_{n+1}) = 0$.

### *5.3.1 An Example of Calculation of* $\pi(\boldsymbol{x}, \boldsymbol{u})$

Let

$$\mathscr{A}_n = \{\boldsymbol{x} \mid T_n(\boldsymbol{x}) = 0\}. \tag{5.30}$$

In general,

$$\pi(\boldsymbol{x}_n, \boldsymbol{u}_n) = \int_{\boldsymbol{y} \in \mathscr{A}_{n+1}} \mathrm{d}P(\boldsymbol{y}, a; \boldsymbol{x}_n, \boldsymbol{u}_n, n). \tag{5.31}$$

For the special case considered in Section 5.2.3, this becomes simply

$$\pi(\boldsymbol{x}_n, \boldsymbol{u}_n) = \sum_{i=1}^{k} p_i \, \delta_i(\boldsymbol{x}_n, \boldsymbol{u}_n), \tag{5.32}$$

with

$$\delta_i(\boldsymbol{x}_n, \boldsymbol{u}_n) = \begin{cases} 1, & T_{n+1}(\boldsymbol{f}^i(\boldsymbol{x}_n, \boldsymbol{u}_n, n)) = 0, \\ 0, & T_{n+1}(\boldsymbol{f}^i(\boldsymbol{x}_n, \boldsymbol{u}_n, n)) \neq 0. \end{cases}$$

### *5.3.2 Recurrence Relation in the Nonstationary Case*

Introducing the probability $\pi(\boldsymbol{x}_n, \boldsymbol{u}_n)$ into (5.15) yields

$$\varphi_n(\boldsymbol{x}_n) = \underset{\boldsymbol{u}_n \in \Omega_n}{\text{opt}} \, \mathrm{E}\{r_n(\boldsymbol{x}_n, \boldsymbol{u}_n) + [1 - \pi_n(\boldsymbol{x}_n, \boldsymbol{u}_n)]\varphi_{n+1}(\boldsymbol{x}_{n+1})\}, \tag{5.33}$$

since the mathematical expectation of the future return between times $n + 1$ and $N$ depends on the probability that the process continues to evolve:

$$(1 - \pi_n) = \text{prob}\{n + 1 \neq N\}. \tag{5.34}$$

This equation is very similar to (5.15). However, since $N$ is unknown, it can not be initialized with

$$\varphi_N(\boldsymbol{x}) = 0.$$

Rather, the only available condition which must be satisfied by the sequence $\varphi_n(\boldsymbol{x})$ is

$$\varphi_n(\boldsymbol{x}) = 0, \qquad \text{if} \quad T_n(\boldsymbol{x}) = 0. \tag{5.35}$$

In general this condition is not sufficient to allow solution of (5.33). However, if there exists an $N$ such that

$$T_n(\boldsymbol{x}) = 0, \qquad \forall \boldsymbol{x} \quad \text{and} \quad n > N,$$

which implies simply that the process will certainly terminate before time $N$, then the relation (5.33) can be solved by initializing with

$$\varphi_N(\boldsymbol{x}) = 0.$$

### 5.3.3 Recurrence Relation in the Stationary Case

If the probability functions defining the problem, $P(\,)$ and $\pi(\,)$, do not depend explicitly on $n$, the system is said to be stationary. In this case the function $\varphi_{n+1}(\boldsymbol{x})$ is identical to $\varphi_n(\boldsymbol{x})$ since the time no longer enters to yield various expected gains, for the same initial state. The index $n$ thus can be dropped from $\varphi$, and (5.33) becomes

$$\varphi(\boldsymbol{x}) = \operatorname*{opt}_{\boldsymbol{u} \in \Omega} \mathrm{E}[r(\boldsymbol{x}, \boldsymbol{u}) + [1 - \pi(\boldsymbol{x}, \boldsymbol{u})]\varphi(\boldsymbol{x}^+)]. \tag{5.36}$$

This is an implicit equation, which can be written

$$\varphi(\boldsymbol{x}) = \operatorname*{opt}_{\boldsymbol{u} \in \Omega} \int_{\mathscr{E}} \{a + [1 - \pi(\boldsymbol{x}, \boldsymbol{u})]\varphi(\boldsymbol{y})\}\, \mathrm{d}P(\boldsymbol{y}, a; \boldsymbol{x}, \boldsymbol{u}). \tag{5.37}$$

In the case considered in Section 5.2.3, the following implicit equation results, with summation replacing integration:

$$\varphi(\boldsymbol{x}) = \operatorname*{opt}_{\boldsymbol{u} \in \Omega} \sum_{i=1}^{k} p_i[r^i(\boldsymbol{x}, \boldsymbol{u}) + (1 - \pi(\boldsymbol{x}, \boldsymbol{u}))\varphi(\boldsymbol{f}^i(\boldsymbol{x}, \boldsymbol{u}))]. \tag{5.38}$$

As in the deterministic case, solution of this implicit equation requires an iterative process.

In general $\varphi(\boldsymbol{x})$ is the limit of the sequence $\psi_n(\boldsymbol{x})$ generated by the recurrence relation

$$\psi_{n+1}(\boldsymbol{x}) = \operatorname*{opt}_{\boldsymbol{u} \in \Omega} \int_{\mathscr{E}} \{a + [1 - \pi(\boldsymbol{x}, \boldsymbol{u})]\psi_n(\boldsymbol{y})\}\, \mathrm{d}P(\boldsymbol{y}, a; \boldsymbol{x}, \boldsymbol{u}), \tag{5.39}$$

with $\psi_0(\boldsymbol{x})$ being some specified initial value, the choice of which will affect the convergence of the process. As in the deterministic case, $\psi_0(\boldsymbol{x})$ can be determined by approximation in either return space or policy space, as discussed in Section 3.4.

## 5.4 Processes with a State Not Completely Measurable

### 5.4.1 Framework of the Problem

For some processes, the state is not directly accessible for measurement. Let us suppose that the only available knowledge about the evolution of the process from time 0 to time $n$ is represented by a data sequence $\boldsymbol{I}_0, \ldots, \boldsymbol{I}_n$, denoted $[\boldsymbol{I}]_0^n$, with each $\boldsymbol{I}_n$ stochastically related to the state vector $\boldsymbol{x}_n$. Thus the control $\boldsymbol{u}_n$ must be determined from the data set $[\boldsymbol{I}]_0^n$.

For an *a priori* control law $\boldsymbol{u}_n, \boldsymbol{u}_{n+1}, \ldots, \boldsymbol{u}_{N-1}$, the return $R_n^N$ realized from time $n$ to time $N$ is a random variable. The optimal policy is the sequence of controls

$$\boldsymbol{u}_n = \mathscr{U}([\boldsymbol{I}]_0^n),$$

which minimizes the mathematical expectation of this return. This expectation should take account of the data set $[\boldsymbol{I}]_0^n$, and thus should be a conditional mathematical expectation, denoted $\mathrm{E}|_0^n[\ ]$, calculated using conditional probability functions.

The process in question is described by the collection of probability functions

$$\begin{aligned}\mathscr{P}(a, \boldsymbol{z}, n; \boldsymbol{I}_0, \ldots, \boldsymbol{I}_n, \boldsymbol{u}) &\\ = \operatorname{prob}\{\boldsymbol{I}_{n+1} \leq \boldsymbol{z} \text{ and } r_n \leq a; \quad &\text{if} \quad [\boldsymbol{I}]_0^n = \boldsymbol{I}_0, \ldots, \boldsymbol{I}_n \\ &\text{and if the control is } \boldsymbol{u}\}. \qquad (5.40)\end{aligned}$$

This distribution function can be calculated from the stochastic model of the process, and the dependence of the measurement $\boldsymbol{I}_n$ on the state, this latter relation being also stochastic.

Making use of (5.40), we have, for example,

$$\mathrm{E}|_0^n[r_n] = \int_{\mathscr{E}} a \, \mathrm{d}\mathscr{P}. \tag{5.41}$$

### *5.4.2 The Optimality Equation*

Just as does the control function, the optimal return function depends on $[\boldsymbol{I}]_0^n$:

$$\operatorname{opt} \mathrm{E}[R_n^N] = \varphi_n^*[\boldsymbol{I}_0, \ldots, \boldsymbol{I}_n].$$

The principle of optimality can be applied to this, yielding

$$\varphi_n^*[\boldsymbol{I}_0, \ldots, \boldsymbol{I}_n] = \operatorname{opt} \ \mathrm{E}|_0^n\{r_n + \varphi_{n+1}^*[\boldsymbol{I}_0, \ldots, \boldsymbol{I}_n, \boldsymbol{I}_{n+1}]\},$$

which is to say

$$\varphi_n^*[\boldsymbol{I}_0, \ldots, \boldsymbol{I}_n] = \operatorname*{opt}_{\boldsymbol{u} \in \Omega} \int_{\mathscr{E}} \{a + \varphi_{n+1}^*[\boldsymbol{I}_0, \ldots, \boldsymbol{I}_n, \boldsymbol{z}]\} \times \mathrm{d}\mathscr{P}[a, \boldsymbol{z}, n; \boldsymbol{I}_0, \ldots, \boldsymbol{I}_n, \boldsymbol{u}]. \tag{5.42}$$

This recurrence relation may be initialized with

$$\varphi_N^*[\boldsymbol{I}_0, \ldots, \boldsymbol{I}_N] \equiv 0. \tag{5.43}$$

Solution of (5.42) allows construction of the optimal control law,

$$\boldsymbol{u}_n = \boldsymbol{g}_n^*(\boldsymbol{I}_0, \ldots, \boldsymbol{I}_n). \tag{5.44}$$

**Remark 1.** In general, $\varphi_n^*(\;)$ and $\boldsymbol{g}_n^*(\;)$ are functions of a number of variables which increases with $n$. From the practical point of view, this severely limits $N$, because of the problem of dimensionality.

**Remark 2.** In practice, the necessity for calculating the functions $\mathscr{P}$ from the data of the problem is a barrier to the use of this method. Only the Gaussian case leads to usable results.

**Remark 3.** It will be shown in Chapter 10 that, if the process is linear and if the perturbations are independent and Gauss–Markov, then $\boldsymbol{I}_0, \ldots, \boldsymbol{I}_n$ can be replaced by a vector $\hat{\boldsymbol{x}}$, which has the same number of dimensions as the state vector $\boldsymbol{x}$, and which is the best estimate of $\boldsymbol{x}$. The functions $\varphi^*(\;)$ and $\boldsymbol{g}^*(\;)$ then become functions of $\hat{\boldsymbol{x}}$, and the difficulties mentioned in Remark 1 disappear. This is a case in which it is possible to extract from $[\boldsymbol{I}\,]_0{}^n$ a complete statistic, such that the sequence $[\boldsymbol{I}\,]_0{}^n$ can be completely reconstructed from a constant number $k$ of quantities, forming a vector $\boldsymbol{X}$ which then plays a role analogous to that of the state vector.

### *5.4.3 Special Case of a Controller without Memory*

For practical reasons, the case sometimes arises in which, at time $n$, only $\boldsymbol{I}_n$ is available for use in determining the optimal control. Then the best control function of the type

$$\boldsymbol{u}_n = \tilde{\boldsymbol{g}}(\boldsymbol{I}_n)$$

is sought. This control should, at time $n$, optimize the mathematical expectation of the return $R_n{}^N$, conditioned only on $\boldsymbol{I}_n$.

The process is, in this case, described by

$$\mathscr{P}(a, z, n; \boldsymbol{I}_n, \boldsymbol{u}).$$

The optimal return $\tilde{\varphi}_n[\;]$ is a function only of $\boldsymbol{I}_n$, so that

$$\tilde{\varphi}_n[\boldsymbol{I}_n] = \operatorname*{opt}_{\boldsymbol{u} \in \Omega} \int_{\mathscr{E}} (a + \tilde{\varphi}_{n+1}[z]) \, d\mathscr{P}(a, z, n; \boldsymbol{I}_n, \boldsymbol{u}). \tag{5.45}$$

This equation is identical to that for the case of an observable state, with $\boldsymbol{I}_n$ playing the role of state vector.

**Remark 1.** Here the calculation of $\mathscr{P}(\,)$ is much easier than in the general case.

**Remark 2.** To get around the difficulties of dimensionality encountered in the general case, one could decide, *a priori*, to replace the sequence $\boldsymbol{I}_0, \ldots, \boldsymbol{I}_n$ by a single quantity $\boldsymbol{I}_n'$, satisfying the recursion

$$\boldsymbol{I}_n' = \mathscr{I}(\boldsymbol{I}_{n-1}', \boldsymbol{I}_n),$$

and to use only $\boldsymbol{I}_n'$ in the calculation of $\boldsymbol{u}_n$. The situation then reverts to that considered in the present section.

The choice of the reduction function $\mathscr{I}(\,)$ directly determines the quality of the controller which results. In any event, in the case of minimization,

$$\tilde{\varphi}_n[\boldsymbol{I}_n'] \geq \varphi_n^*[\boldsymbol{I}_0, \ldots, \boldsymbol{I}_n] \geq \varphi_n(\boldsymbol{x}_n), \tag{5.46}$$

which simply states that the more information that is available about the process, the better will be the control. An optimal estimate of the state $\boldsymbol{x}_n$ is sometimes used for $\mathscr{I}(\,)$, such as its conditional mathematical expectation, or its most probable value, even if the estimate used is not a complete statistic for the problem in question.

## 5.5 Conclusions

We have seen that for a stochastic process only a closed-loop control is acceptable, and that the optimum control can be defined only in terms of the average of the future return.

With use of dynamic programming, the theoretical solution to this control problem can easily be found. Its structure is analogous to that obtained in the deterministic case, namely, a functional recursion relation in the case of a bounded horizon, and an implicit equation in the case of a horizon which is not fixed in advance.

On the practical level, solution of these equations requires additional calculations for evaluation of the mathematical expectation. In case the state is measurable, numerical methods patterned after those developed in Chapter 4 can be used.

The more realistic case is that in which the state is not exactly measurable, and here the solution is difficult to carry out, mainly because of the increasing number of arguments of the functions involved. The curse of dimensionality then limits $N$ severely, which removes most of the practical interest in this approach. In the rare cases in which a complete statistic exists, the problem can be stated in terms of functions with a constant number of arguments, and an optimal solution obtained for any $N$. If this is not the case, the optimality of the solution can be sacrificed, and an *a priori* method used to reduce the amount of information about the past states. With this method, some approximation to the optimal solution can be found.

# Chapter 6 Processes with Discrete States

In this chapter we shall study the optimal control of discrete stochastic processes having discrete states. After a discussion of the definitions and properties of this class of systems, we shall consider two types of problems, those in which only a terminal cost is present, and those in which the cost criterion is determined by elementary returns.

Since we are treating a simple case of the general problem studied in the preceding chapter, we shall be led to particularly simple numerical methods for its solution. The simplification comes about because the discrete character of the states allows the mathematical expectations to be expressed in terms of the state transition probabilities.

Discrete-state processes are often used to model systems defined only qualitatively, such as economic, social, or game processes. In addition, all digital systems are of this type, if some chance switching is introduced. Further, quantizing the variables of a continuous-state process leads to an approximate model of the type to be considered here.

## 6.1 Fundamentals

### *6.1.1 Definition of a Discrete Process with Discrete States*

If the state is discrete, at each instant it can be represented by a number $i$:

$$e_n \in \{1, \ldots, i, \ldots, M\}, \tag{6.1}$$

where $M$ is the total number of possible states, which we shall assume to be finite.

Passage from the state $i$, at time $n$, to the state $j$ at time $n+1$, is a random event, which we shall assume depends only on the state $i$, the time $n$, and a control $\boldsymbol{u}$. A transition probability can then be defined:

$$P_{ij}(\boldsymbol{u}, n) = \text{prob}\{e_{n+1} = j \quad \text{if} \quad e_n = i \quad \text{and} \quad \text{control} = \boldsymbol{u}\}. \tag{6.2}$$

The applied control $\boldsymbol{u}$ can be a member of either a continuous or a discrete ensemble of possible controls. In the latter case it will be represented by a number $k$. The model thus defined is also called a finite-state automaton, with random evolution. The definition of a discrete Markov process with discrete states, presented by Boudarel *et al.* [1, Vol. 1, Chap. 13], can also be recognized.

In order that they indeed be probabilities, the functions $P_{ij}(\boldsymbol{u}, n)$ must satisfy certain conditions, in particular,

$$0 \leq P_{ij}(\boldsymbol{u}, n) \leq 1, \tag{6.3}$$

$$\sum_{j=1}^{M} P_{ij}(\boldsymbol{u}, n) = 1, \qquad \forall i. \tag{6.4}$$

In the following, the $P_{ij}(\boldsymbol{u}, n)$ will be considered to be the elements of a matrix $\mathbf{P}(\boldsymbol{u}, n)$.

### 6.1.2 *A Priori Probabilities and the Stochastic Steady State*

If at the initial time the system is in a known state $i_0$, the state attained at time $n$, using the policy $\boldsymbol{u}_0, \ldots, \boldsymbol{u}_{n-1}$, is random. However, the probability of being in any particular state at time $n$ can be calculated. Let us introduce a vector of probabilities $\boldsymbol{v}_n$, with

$$v_n^j = \text{prob}\{\text{state} = j \quad \text{at time} \quad n\}. \tag{6.5}$$

The vector $\boldsymbol{v}_n$ should not be confused with the deterministic state vector.

If the initial state is also random, with a vector $\boldsymbol{v}_0$ of *a priori* probabilities, it follows by application of the composition law for probabilities, that

$$\boldsymbol{v}_n = \mathbf{P}^{\mathrm{T}}(\boldsymbol{u}_{n-1}, n-1) \cdots \mathbf{P}^{\mathrm{T}}(\boldsymbol{u}_0, 0)\boldsymbol{v}_0, \tag{6.6}$$

since, for any $n$,

$$\boldsymbol{v}_{n+1} = \mathbf{P}^{\mathrm{T}}(\boldsymbol{u}_n, n)\boldsymbol{v}_n.$$

In case $\mathbf{P}(\boldsymbol{u}_n, n)$ is constant, (6.6) can be written

$$\boldsymbol{v}_n = (\mathbf{P}^{\mathrm{T}})^n \boldsymbol{v}_0. \tag{6.7}$$

As discussed by Boudarel *et al.* [1, Vol. 1, p. 228], if the spectrum of the matrix $\mathbf{P}$ is such that the only eigenvalue of $\mathbf{P}$ with modulus one is unity, with multiplicity one, then when $n$ tends to infinity, $\boldsymbol{v}_n$ tends to a vector $\boldsymbol{v}$, which is independent of $\boldsymbol{v}_0$ and which satisfies

$$\bar{\boldsymbol{v}} = \mathbf{P}^{\mathrm{T}}\bar{\boldsymbol{v}} \tag{6.8}$$

with

$$\bar{\boldsymbol{v}}^{\mathrm{T}}\boldsymbol{1} = 1 .$$

The matrix $\mathbf{P}$ is then said to be ergodic.

When the matrices $\mathbf{P}(\boldsymbol{u}_m, m)$ are periodic beyond some value $m'$, with period $p$, that is,

$$\mathbf{P}(\boldsymbol{u}_{n+p}, n+p) = \mathbf{P}(\boldsymbol{u}_n, n) = \mathbf{P}_l, \tag{6.9}$$

where

$$n > m', \qquad l = [n - m']_{\mathrm{mod}\,p},$$

a periodic steady-state regime is possible if the matrices

$$\mathbf{\Pi}_l^{\mathrm{T}} = \mathbf{P}_{l-1}^{\mathrm{T}}\mathbf{P}_{l-2}^{\mathrm{T}} \cdots \mathbf{P}_1^{\mathrm{T}}\mathbf{P}_p^{\mathrm{T}}\mathbf{P}_{p-1}^{\mathrm{T}} \cdots \mathbf{P}_l^{\mathrm{T}} \tag{6.10}$$

are all ergodic. Observation of the state once each $p$ time instants then leads to a constant probability vector, which is independent of the initial state, but which can differ depending on the phase of the observations. Letting $\boldsymbol{v}_l$ be such that

$$\bar{\boldsymbol{v}}_l^{\mathrm{T}}\boldsymbol{1} = 1, \qquad \bar{\boldsymbol{v}}_l = \mathbf{\Pi}^{\mathrm{T}}\bar{\boldsymbol{v}}_l, \tag{6.11}$$

the periodic steady state will be defined by the sequences

$$\text{control} \to \boldsymbol{u}_{p-1}, \boldsymbol{u}_p, \boldsymbol{u}_1, \boldsymbol{u}_2, \ldots, \boldsymbol{u}_l, \ldots, \boldsymbol{u}_p, \boldsymbol{u}_1, \ldots$$

$$\bar{\boldsymbol{v}} \to \bar{\boldsymbol{v}}_{p-1}, \bar{\boldsymbol{v}}_p, \bar{\boldsymbol{v}}_1, \bar{\boldsymbol{v}}_2, \ldots, \bar{\boldsymbol{v}}_l, \ldots, \bar{\boldsymbol{v}}_p, \bar{\boldsymbol{v}}_1, \ldots .$$

### 6.1.3 *Example*

Let us consider a stationary process with two states, and with a discrete control which can take two values. The system is diagrammed in Fig. 6.1. Depending on the control, there are two transition matrices,

$$\mathbf{P}(1) = \begin{bmatrix} 0.3 & 0.7 \\ 0.9 & 0.1 \end{bmatrix}, \qquad \mathbf{P}(2) = \begin{bmatrix} 0.6 & 0.4 \\ 0.2 & 0.8 \end{bmatrix}.$$

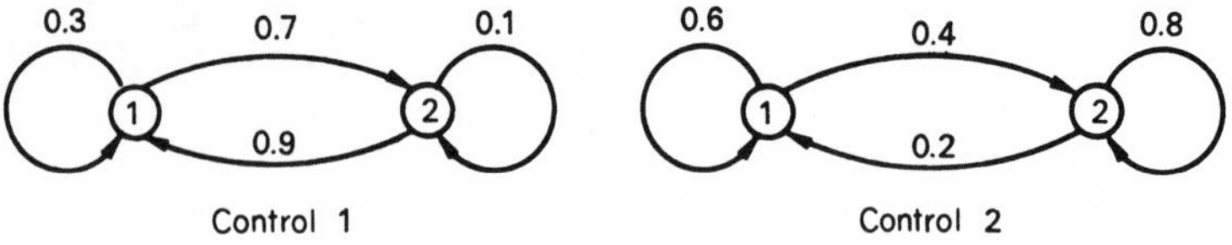


FIG. 6.1. Transition probabilities for a discrete-state process.

According to (6.8), steady application of control 1 leads to a steady state such that

$$\begin{bmatrix} 0.3 & 0.9 \\ 0.7 & 0.1 \end{bmatrix} \begin{bmatrix} x \\ 1-x \end{bmatrix} = \begin{bmatrix} x \\ 1-x \end{bmatrix} \Rightarrow \bar{\boldsymbol{v}}_1 = \begin{bmatrix} \frac{9}{16} \\ \frac{7}{16} \end{bmatrix}$$

while control 2 leads to

$$\begin{bmatrix} 0.6 & 0.2 \\ 0.4 & 0.8 \end{bmatrix} \begin{bmatrix} x \\ 1-x \end{bmatrix} = \begin{bmatrix} x \\ 1-x \end{bmatrix} \Rightarrow \bar{\boldsymbol{v}}_2 = \begin{bmatrix} \frac{1}{3} \\ \frac{2}{3} \end{bmatrix}.$$

Note that control 1 leads to a greater probability of being in state 1 in the steady state than does control 2.

Use of the periodic control sequence 1, 2, 1, 2, ... leads to a steady state with period two. According to (6.10) and (6.11),

$$\boldsymbol{\Pi}_1{}^{\mathrm{T}} = \mathbf{P}^{\mathrm{T}}(2)\mathbf{P}^{\mathrm{T}}(1) = \begin{bmatrix} 0.32 & 0.56 \\ 0.68 & 0.44 \end{bmatrix} \Rightarrow \bar{\boldsymbol{v}}' = \begin{bmatrix} \frac{14}{31} \\ \frac{17}{31} \end{bmatrix},$$

$$\boldsymbol{\Pi}_2{}^{\mathrm{T}} = \mathbf{P}^{\mathrm{T}}(1)\mathbf{P}^{\mathrm{T}}(2) = \begin{bmatrix} 0.54 & 0.78 \\ 0.46 & 0.22 \end{bmatrix} \Rightarrow \bar{\boldsymbol{v}}'' = \begin{bmatrix} \frac{39}{62} \\ \frac{23}{62} \end{bmatrix}.$$

Note that the largest probability of being in state 1, $\frac{39}{62}$, is obtained at the time of application of control 2, and that this exceeds the probability $\frac{9}{16}$ which resulted from steady application of control 1.

## 6.2 Terminal Problems

We designate as terminal problems those in which the probability of being in a given state at a given instant is to be optimized. More particularly, we shall study problems in which the final time $N$ is given, and then problems in which $N$ is a random variable.

### 6.2.1 *Horizon Fixed in Advance*

To simplify the development, we shall suppose that the final state is $j = 1$, and that the matrix $\mathbf{P}$ is independent of $n$ (stationary evolution). Then let $\hat{w}_i(n)$ be the probability that the system will be in state 1 at time $N$ if it is in state $i$ at time $n$, and if the optimal policy is used. Use of any control $\boldsymbol{u}$ at time $n$ results in a probability $P_{ij}(\boldsymbol{u})$ of being in state $j$ at time $n + 1$. For this state, there is a probability $\hat{w}_j(n + 1)$ of terminating in state 1 at time $N$ (since the optimal policy is used from that point on).

With these considerations, the formula for composition of probabilities and the principle of optimality yield

$$\hat{w}_i(n) = \underset{\boldsymbol{u}}{\text{opt}} \sum_{j=1}^{M} P_{ij}(\boldsymbol{u})\, \hat{w}_j(n + 1),$$

or in matrix form

$$\hat{\mathbf{w}}(n) = \underset{\boldsymbol{u}}{\text{opt}}[\mathbf{P}(\boldsymbol{u})\hat{\mathbf{w}}(n + 1)]. \tag{6.12}$$

This is a recurrence relation among the $\hat{\mathbf{w}}(n)$. It is to be initialized with

$$\hat{\mathbf{w}}(N) = \begin{bmatrix} 1 \\ 0 \\ \vdots \\ 0 \end{bmatrix}.$$

**Remark 1.** If time is measured backward from the final time, (6.12) should be written

$$\hat{\mathbf{w}}_{m+1} = \underset{\boldsymbol{u}}{\text{opt}}[\mathbf{P}(\boldsymbol{u})\hat{\mathbf{w}}_m], \qquad \hat{\mathbf{w}}_0 = \begin{bmatrix} 1 \\ 0 \\ \vdots \\ 0 \end{bmatrix}. \tag{6.13}$$

**Remark 2.** For each $i$, minimization of the bracketed quantity in (6.12) yields a control $\hat{\boldsymbol{u}}$, which allows construction of the optimal control function $\hat{\boldsymbol{u}} = \boldsymbol{g}_n(i)$. This can be represented as a vector $\boldsymbol{U}_n$, with

$$\boldsymbol{U}_n^{\,i} = \boldsymbol{g}_n(i). \tag{6.14}$$

### 6.2.2 *Example*

Let us consider a process with two states, 1 and 2, a control $u$ with $0.2 \le u \le 0.8$, and a transition matrix

$$\mathbf{P}(u) = \begin{bmatrix} 0.4u & 1 - 0.4u \\ 1 - u^2 & u^2 \end{bmatrix}.$$

It is easy to verify that

$$\sum_{j=1}^{2} P_{ij}(u) = 1, \qquad P_{ij}(u) \ge 0, \quad \forall \text{ admissible } u.$$

The probability of being in state 1 at the final time is to be maximized. Counting time backward from the final time, we have

$$\hat{w}_0 = \begin{bmatrix} 1 \\ 0 \end{bmatrix}$$

$$\hat{w}_1 = \max_u \begin{bmatrix} 0.4u \\ 1 - u^2 \end{bmatrix} = \begin{bmatrix} 0.32 \\ 0.96 \end{bmatrix} \Leftrightarrow \boldsymbol{U}_1 = \begin{bmatrix} 0.8 \\ 0.2 \end{bmatrix}$$

$$\hat{w}_2 = \max_u \begin{bmatrix} 0.32 \times 0.4u + (1 - 0.4u)0.96 \\ 0.32(1 - u^2) + 0.96u^2 \end{bmatrix} = \begin{bmatrix} 0.909 \\ 0.73 \end{bmatrix} \Leftrightarrow \boldsymbol{U}_2 = \begin{bmatrix} 0.2 \\ 0.8 \end{bmatrix}$$

$$\hat{w}_3 = \max_u \begin{bmatrix} 0.909 \times 0.4u + (1 - 0.4u)0.73 \\ 0.909(1 - u^2) + 0.73u^2 \end{bmatrix} = \begin{bmatrix} 0.789 \\ 0.895 \end{bmatrix} \Leftrightarrow \boldsymbol{U}_3 = \begin{bmatrix} 0.8 \\ 0.2 \end{bmatrix}$$

$$\hat{w}_4 = \max_u \begin{bmatrix} 0.789 \times 0.4u + (1 - 0.4u)0.895 \\ 0.789(1 - u^2) + 0.895u^2 \end{bmatrix} = \begin{bmatrix} 0.8865 \\ 0.857 \end{bmatrix} \Leftrightarrow \boldsymbol{U}_4 = \begin{bmatrix} 0.2 \\ 0.8 \end{bmatrix}$$

$$\hat{w}_5 = \max_u \begin{bmatrix} 0.8865 \times 0.4u + (1 - 0.4u)0.857 \\ 0.8865(1 - u^2) + 0.857u^2 \end{bmatrix} = \begin{bmatrix} 0.8664 \\ 0.8852 \end{bmatrix} \Leftrightarrow \boldsymbol{U}_5 = \begin{bmatrix} 0.8 \\ 0.2 \end{bmatrix}.$$

The optimal control law is summarized in Table 6.1.

**Table 6.1**

| | $m$ | 5 | 4 | 3 | 2 | 1 | Final time ↓ |
|---|---|---|---|---|---|---|---|
| Control | If state 1 → | 0.8 | 0.2 | 0.8 | 0.2 | 0.8 | |
| | If state 2 → | 0.2 | 0.8 | 0.2 | 0.8 | 0.2 | |

The existence of a periodic regime can be verified for the control:

$$m \text{ even}\begin{cases}\text{state } 1 \to \hat{u} = 0.2\\ \text{state } 2 \to \hat{u} = 0.8\end{cases} \to \mathbf{P}(\hat{u}) = \mathbf{P}_2 = \begin{bmatrix}0.08 & 0.92\\ 0.36 & 0.64\end{bmatrix}$$

$$m \text{ odd}\begin{cases}\text{state } 1 \to \hat{u} = 0.8\\ \text{state } 2 \to \hat{u} = 0.2\end{cases} \to \mathbf{P}(\hat{u}) = \mathbf{P}_1 = \begin{bmatrix}0.32 & 0.68\\ 0.96 & 0.04\end{bmatrix}.$$

Let us seek the limit $\bar{\boldsymbol{v}}$ in phase with the final time. This corresponds to a transition matrix $\boldsymbol{\Pi}_2$ such that

$$\boldsymbol{\Pi}_2{}^{\mathrm{T}} = \mathbf{P}_1{}^{\mathrm{T}}\mathbf{P}_2{}^{\mathrm{T}} = \begin{bmatrix}0.9088 & 0.7296\\ 0.0912 & 0.2704\end{bmatrix}.$$

This matrix is ergodic, and leads to

$$\bar{\boldsymbol{v}} = \begin{bmatrix}0.89\\ 0.11\end{bmatrix}.$$

Thus, for large $N$, the probability of terminating in state 1 is 0.89 whatever the initial state. This result is in good agreement with that obtained by calculation of the $\hat{\boldsymbol{w}}_n$.

### 6.2.3 *Random Horizon*

We shall now examine the case that each instant $n$ has a probability $q$ of being the final time. The horizon, or stopping time, $N$ is thus a discrete Poisson process.

Let $\hat{\boldsymbol{w}}$ have components $\hat{w}_i$ which are the probabilities of terminating in state 1 if the initial state is $i$, and if an optimal control is used.

It is evident that $\hat{\boldsymbol{w}}$ does not depend on $n$. The principle of optimality then yields

$$\hat{w}_i = \underset{\boldsymbol{u}}{\mathrm{opt}}\left[qP_{i1}(\boldsymbol{u}) + (1-q)\sum_{j=1}^{M} P_{ij}(\boldsymbol{u})\hat{w}_j\right]. \qquad (6.15)$$

The first term is the probability that the following step reaches state 1 and that the process then terminates, while the second term is the probability that state 1 is reached at some step beyond the following step, using the optimal policy.

The implicit relation (6.15) is solved by an iterative process. From the theoretical point of view, this raises the problem of the existence of a solution. We shall return to this point in Section 6.3, where an equation analogous to (6.15) is encountered.

### 6.2.4 *Example*

Let us continue the preceding example, using $q = 0.2$. Equation (6.15) leads to

$$\hat{w}_1 = \max_u [0.2(0.4u) + 0.8(0.4u\hat{w}_1 + (1 - 0.4u)\hat{w}_2)]$$
$$\hat{w}_2 = \max_u [0.2(1 - u^2) + 0.8((1 - u^2)\hat{w}_1 + u^2\hat{w}_2)].$$

Solving this system iteratively, using the initial values

$$\hat{w}^0 = \begin{bmatrix} 0.5 \\ 0.5 \end{bmatrix},$$

leads to the results given in Table 6.2.

**Table 6.2**

| | $n=1$ | $n=2$ | $n=3$ | $n=4$ | $n=5$ | $n=6$ | $n=7$ |
|---|---|---|---|---|---|---|---|
| $\hat{w}_1$ | 0.4640 | 0.5048 | 0.4998 | 0.5170 | 0.5200 | 0.5219 | 0.5315 |
| $\hat{w}_2$ | 0.5920 | 0.5637 | 0.5977 | 0.5950 | 0.6081 | 0.6109 | 0.6170 |
| $U_1$ | 0.8 | 0.8 | 0.8 | 0.8 | 0.8 | 0.8 | 0.8 |
| $U_2$ | 0.2 | 0.2 | 0.2 | 0.2 | 0.2 | 0.2 | 0.2 |

From the table it can be seen that from the very beginning the policy converges to

$$\text{State } 1 \rightarrow u = 0.8;$$
$$\text{State } 2 \rightarrow u = 0.2.$$

For this stationary policy, the probabilities $\hat{w}_1$ and $\hat{w}_2$ of terminating in state 1 satisfy a relation analogous to (6.15), but without the optimization operator, and with $u = 0.8$ in the first equation, and $u = 0.2$ in the second. We thus have the linear system

$$\tilde{w}_1 = 0.2(0.4 \times 0.8) + 0.8[(0.4 \times 0.8)\tilde{w}_1 + (1 - 0.4 \times 0.8)\tilde{w}_2]$$
$$\tilde{w}_2 = 0.2(1 - 0.2 \times 0.2) + 0.8[(1 - 0.2 \times 0.2)\tilde{w}_1 + (0.2 \times 0.2)\tilde{w}_2)]$$

with solution

$$\tilde{\boldsymbol{w}} = \begin{bmatrix} 0.5455 \\ 0.6301 \end{bmatrix},$$

which is indeed the value obtained for $\hat{\boldsymbol{w}}$ by iteration.

The steady state corresponding to the limit policy yields

$$\bar{\boldsymbol{v}} = \begin{bmatrix} 0.586 \\ 0.414 \end{bmatrix}.$$

Thus at any time in the steady state, the probability of being in state 1 is 0.586, which is just the average of $\hat{w}_1$ and $\hat{w}_2$.

## 6.3 Optimization of a Return Function

### *6.3.1 The Problem*

We shall now consider a stationary stochastic process with discrete controls. Such a process can be described by the probabilities $P_{ij}(k)$ that the system goes from state $i$ to state $j$ upon application of control $k$. The total return is the sum of the elementary returns $r_{ij}(k)$, associated with passage from state $i$ to state $j$ under the action of control $k$. In addition, we distinguish two classes of states from the totality of all states: (a) initial states, numbering $M$, such that $1 \leq i \leq M$; (b) terminal states, numbering $m$, such that $M + 1 \leq i \leq M + m$.

The problem is then to go from some one of the initial states to any one of the final states, in such a way that the total return $R$, the sum of the elementary returns, is optimized, the sum being taken over all steps of the process up till the attainment of some one of the final states. Since the evolution of the process is random, however, only the control law which optimizes the mathematical expectation of the total return can be found.

Since the process is stationary and the horizon random (*a priori*), the optimal policy, if it exists, must be stationary. Thus, to each state there will correspond an optimal control to be applied. Introducing a vector $\boldsymbol{U}$ with components $U_i$, where $U_i$ is the number of the control to be applied if the system is in state $i$, the problem can be reduced to a search among the allowable vectors $\boldsymbol{U}$ for that which optimizes the mathematical expectation of the return.

### 6.3.2 *Calculation of the Average Return for an Arbitrary Policy*

Let $\mathbf{y}(\boldsymbol{U})$ have components $y_i$, $i = 1, \ldots, M$, with $y_i$ the mathematical expectation of the return starting from state $i$ and using the control policy $\boldsymbol{U}$. This expected return is the sum of the immediate gain $r_{ij}(U_i)$, if transition to state $j$ occurs, plus the expected gain $y_j(\boldsymbol{U})$ to be realized from the following state $j$, provided $j$ is not a terminal state. Since the transition from state $i$ to some state $j$ occurs with probability $P_{ij}(\boldsymbol{U}_i)$, we thus have

$$y_i(\boldsymbol{U}) = \sum_{j=1}^{M+m} P_{ij}(U_i)r_{ij}(U_i) + \sum_{j=1}^{M} P_{ij}(U_i)y_j. \tag{6.16}$$

Letting $\mathbf{c}(\boldsymbol{U})$ be the vector with components

$$c_i(\boldsymbol{U}) = \sum_{j=1}^{M+m} P_{ij}(U_i)r_{ij}(U_i), \tag{6.17}$$

the expected return (6.16) can be written

$$\mathbf{y}(\boldsymbol{U}) = \mathbf{c}(\boldsymbol{U}) + \mathbf{P}'(\boldsymbol{U})\,\mathbf{y}(\boldsymbol{U}), \tag{6.18}$$

where $\mathbf{P}'$ corresponds to transitions into the interior of the domain of initial states.

If the matrix $\mathbf{1} - \mathbf{P}'(\boldsymbol{U})$ has an inverse, the policy $\boldsymbol{U}$ will be said to be admissible. In this case, the expected return is

$$\mathbf{y}(\boldsymbol{U}) = [\mathbf{1} - \mathbf{P}'(\boldsymbol{U})]^{-1}\mathbf{c}(\boldsymbol{U}). \tag{6.19}$$

### 6.3.3 *Example*

Consider a system with three states, 1, 2, and 3. Let 1 and 2 be initial states, so that $M = 2$, and let the terminal state be 3, so that $m = 1$. Let there be two possible controls, $A$ and $B$, and let the transition probability matrices be

$$\text{control} = A\text{: } \mathbf{P}_A = \begin{bmatrix} 0.1 & 0.1 & 0.8 \\ 0.5 & 0.2 & 0.3 \\ 1.0 & 0 & 0 \end{bmatrix},$$

$$\text{control} = B\text{: } \mathbf{P}_B = \begin{bmatrix} 0.4 & 0.3 & 0.3 \\ 0.1 & 0.8 & 0.1 \\ 1.0 & 0 & 0 \end{bmatrix}.$$

The process is diagrammed in Fig. 6.2.

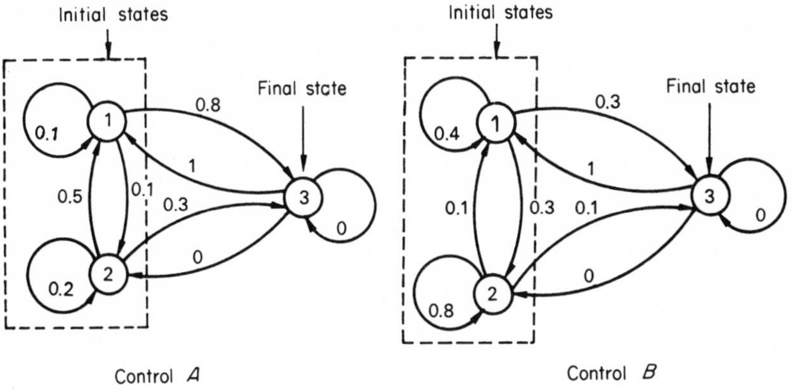


FIG. 6.2. Transition probabilities for a discrete-state process.

It is desired to minimize the expected number of state transitions which occur before attaining the terminal state 3. This leads to the elementary returns

$$r_{ij} = 1, \qquad i \neq j$$
$$r_{ij} = 0, \qquad i = j$$

so that for either of the two possible controls

$$\text{control} = A \text{ or } B\text{: } \mathbf{r} = \begin{bmatrix} 0 & 1 & 1 \\ 1 & 0 & 1 \\ 1 & 1 & 0 \end{bmatrix}.$$

Four different control policies are possible:

$$\begin{bmatrix} A \\ A \end{bmatrix}, \quad \begin{bmatrix} A \\ B \end{bmatrix}, \quad \begin{bmatrix} B \\ B \end{bmatrix}, \quad \begin{bmatrix} B \\ A \end{bmatrix}.$$

The corresponding returns, calculated from (6.19), are shown in Table 6.3. It can be seen that, for this example, there exists a policy which ensures the smallest $y_i$, namely, the policy

$$\begin{bmatrix} A \\ B \end{bmatrix}.$$

### *6.3.4 The Equation of Optimality*

Let $\hat{\mathbf{y}}$ be the optimal return, obtained by using the optimal policy $\hat{\mathbf{U}}$. The principle of optimality yields

$$\hat{y}_i = \operatorname*{opt}_k \left[ c_i(k) + \sum_{j=1}^{M} P_{ij}(k) \hat{y}_j \right], \tag{6.20}$$

**Table 6.3**

| $U$ | $c(U)$ | $\mathbf{P}'(U)$ | $\mathbf{1}-\mathbf{P}'(U)$ | $(\mathbf{1}-\mathbf{P}'(U))^{-1}$ | $y(U)$ |
|---|---|---|---|---|---|
| $A$ | 0.9 | 0.1 0.1 | 0.9 −0.1 | 1.34 0.149 | 1.192 |
| $A$ | 0.8 | 0.5 0.2 | −0.5 0.8 | 0.746 1.19 | 1.75 |
| $A$ | 0.9 | 0.1 0.1 | 0.9 −0.1 | 1.18 0.59 | 1.176 |
| $B$ | 0.2 | 0.1 0.8 | −0.1 0.2 | 0.59 5.4 | 1.59 |
| $B$ | 0.6 | 0.4 0.3 | 0.6 −0.3 | 0.222 0.333 | 2 |
| $B$ | 0.2 | 0.1 0.8 | −0.1 0.2 | 0.111 0.663 | 2 |
| $B$ | 0.6 | 0.4 0.3 | 0.6 −0.3 | 2.42 0.91 | **2.14** |
| $A$ | 0.8 | 0.5 0.2 | −0.5 0.8 | 1.52 1.82 | **2.36** |

where $c_i(k)$ is the mathematical expectation of the immediate return if control $k$ is used, and $\sum P_{ij}(k)\hat{y}_j$ is the mathematical expectation of the future return, which is assumed optimal. In matrix form, (6.20) becomes

$$\hat{\mathbf{y}} = \underset{\mathbf{U}}{\text{opt}}[\mathbf{c}(\mathbf{U}) + \mathbf{P}'(\mathbf{U})\hat{\mathbf{y}}], \tag{6.21}$$

which is an implicit equation, as always if the horizon is not specified in advance.

### *6.3.5 Existence of an Optimal Solution*

The existence of a control policy $\hat{\mathbf{U}}$ such that

$$\mathbf{y}(\hat{\mathbf{U}}) \leq \mathbf{y}(\mathbf{U}), \qquad \forall \text{ admissible } \mathbf{U}$$

is not immediately evident, since the vector inequality

$$\boldsymbol{A} \leq \boldsymbol{B},$$

which means $A_i \leq B_i$ for all $i$, is not a relation of total ordering. Thus the set of the $\mathbf{y}(\mathbf{U})$ does not necessarily contain an "optimal" element.

However, if an optimal element exists, it must satisfy (6.21). It is thus sufficient to demonstrate the existence of a solution of (6.21). As discussed in Section 3.4.2, for this it suffices to show that the operator

$$T = \text{opt}[\ ]$$

is a contraction.

Consider two vectors $\boldsymbol{x}_1$ and $\boldsymbol{x}_2$. We have

$$\begin{aligned} T(\boldsymbol{x}_1) &= \operatorname*{opt}_{\boldsymbol{U}}[\boldsymbol{c}(\boldsymbol{U}) + \mathbf{P}'(\boldsymbol{U})\boldsymbol{x}_1] = \boldsymbol{c}(\boldsymbol{U}_1) + \mathbf{P}'(\boldsymbol{U}_1)\boldsymbol{x}_1, \\ T(\boldsymbol{x}_2) &= \operatorname*{opt}_{\boldsymbol{U}}[\boldsymbol{c}(\boldsymbol{U}) + \mathbf{P}'(\boldsymbol{U})\boldsymbol{x}_2] = \boldsymbol{c}(\boldsymbol{U}_2) + \mathbf{P}'(\boldsymbol{U}_2)\boldsymbol{x}_2. \end{aligned} \tag{6.22}$$

In the case of search for a minimum,

$$\begin{aligned} T(\boldsymbol{x}_1) &\le \boldsymbol{c}(\boldsymbol{U}_2) + \mathbf{P}'(\boldsymbol{U}_2)\boldsymbol{x}_1, \\ T(\boldsymbol{x}_2) &\le \boldsymbol{c}(\boldsymbol{U}_1) + \mathbf{P}'(\boldsymbol{U}_1)\boldsymbol{x}_2. \end{aligned}$$

From these relations there follows

$$\mathbf{P}'(\boldsymbol{U}_1)[\boldsymbol{x}_1 - \boldsymbol{x}_2] \le T(\boldsymbol{x}_1) - T(\boldsymbol{x}_2) \le \mathbf{P}'(\boldsymbol{U}_2)[\boldsymbol{x}_1 - \boldsymbol{x}_2],$$

and hence

$$\begin{aligned} |T(\boldsymbol{x}_1) - T(\boldsymbol{x}_2)| &\le \max[|\mathbf{P}'(\boldsymbol{U}_1)(\boldsymbol{x}_1 - \boldsymbol{x}_2)|, |\mathbf{P}'(\boldsymbol{U}_2)(\boldsymbol{x}_1 - \boldsymbol{x}_2)|] \\ &\le \max_{\boldsymbol{U}} |\mathbf{P}'(\boldsymbol{U})(\boldsymbol{x}_1 - \boldsymbol{x}_2)|. \end{aligned}$$

Defining as the norm of the vector $\boldsymbol{x}$,

$$\|\boldsymbol{x}\| = \max_i |x_i|,$$

we then have

$$\begin{aligned} \|T(\boldsymbol{x}_1) - T(\boldsymbol{x}_2)\| &\le \max_i \max_{\boldsymbol{U}} \left| \sum_{j=1}^{M} P_{ij}(\boldsymbol{U})(x_1{}^j - x_2{}^j) \right| \\ &\le \max_i \max_{\boldsymbol{U}} \sum_{j=1}^{M} |P_{ij}(\boldsymbol{U})| \, \|\boldsymbol{x}_1 - \boldsymbol{x}_2\|. \end{aligned}$$

Letting

$$a = \max_i \max_{\boldsymbol{U}} \sum_{j=1}^{M} |P_{ij}(\boldsymbol{U})|,$$

and taking account of the fact that

$$\forall \boldsymbol{U} : \sum_{j=1}^{M+m} P_{ij} = 1, \qquad P_{ij} > 0 \Rightarrow a \leq 1,$$

we obtain finally

$$\| T(\boldsymbol{x}_1) - T(\boldsymbol{x}_2) \| = a \, \| \boldsymbol{x}_1 - \boldsymbol{x}_2 \| .$$

Since the space of the $\boldsymbol{x}$ is denumerable, it is complete, and using the contraction theorem, we have that if

$$a = \max_{i} \max_{U} \sum_{j=1}^{m} | P_{ij}(U) | < 1, \tag{6.23}$$

then Eq. (6.21) has a unique solution, which can be obtained iteratively starting from any initial value. This result is only a necessary condition for existence of an optimal control policy.

Applying this result to the preceding example, we have

| | $A$ | $B$ |
|---|---|---|
| $P_{11} + P_{12}$ | 0.2 | 0.7 |
| $P_{21} + P_{22}$ | 0.7 | **0.9** |

which leads to $a = 0.9 < 1$, so that a unique solution exists.

### 6.3.6 *Practical Solution*

When the condition of Eq. (6.23) has been satisfied, a method of successive approximations can be used for the solution of (6.21). As soon as convergence toward a policy is observed, the return corresponding to that policy can be calculated directly from (6.19). This allows the final result to be obtained without waiting for final convergence. However, it is then necessary to solve a linear system in $M$ unknowns, which can involve numerical difficulties if $M$ is large. Figure 6.3 shows a general flow chart for solution of (6.21), using successive approximations.

### 6.3.7 *Example*

Returning again to the above example, application of (6.20) yields

$$y_1 = \min[0.1y_1 + 0.1y_2 + 0.9,\ 0.4y_1 + 0.3y_2 + 0.6],$$

$$y_2 = \min[0.5y_1 + 0.2y_2 + 0.7,\ 0.1y_1 + 0.8y_2 + 0.2].$$

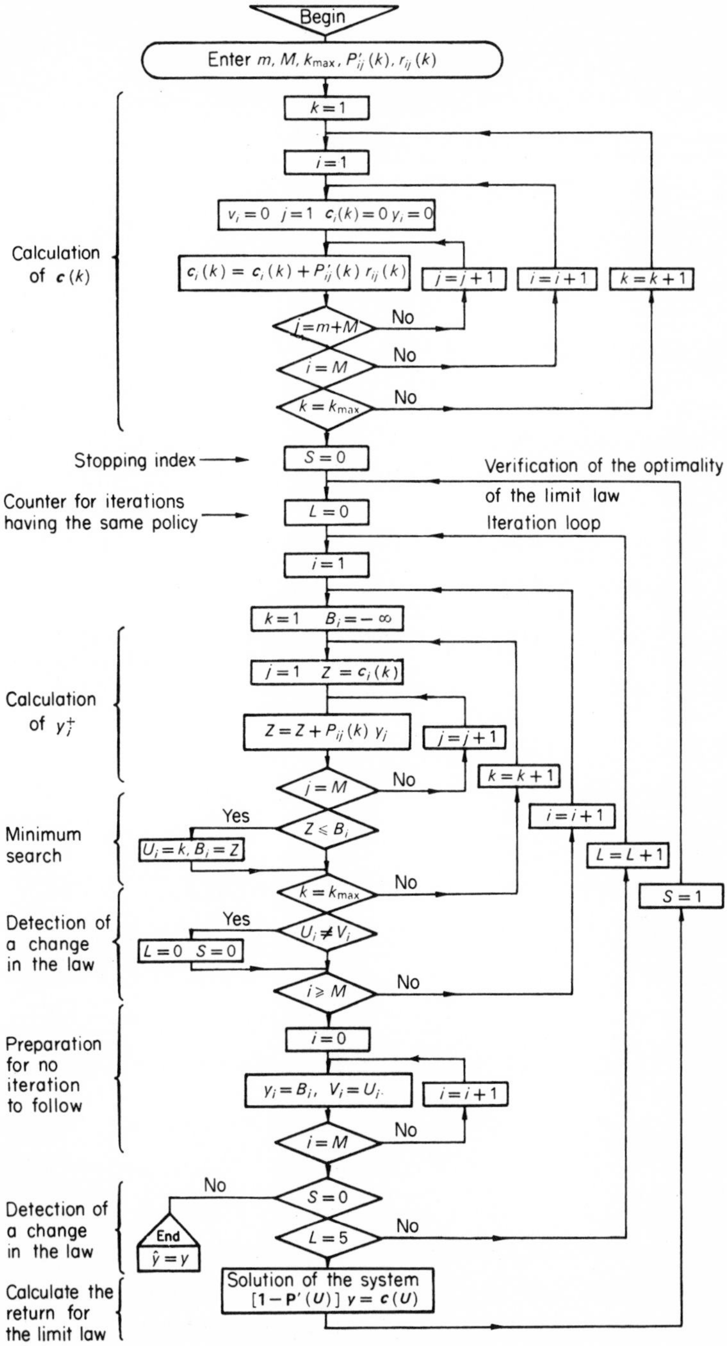


FIG. 6.3. Iterative computation of the solution to Equation (6.21).

The iterative solution, starting from

$$y = \begin{bmatrix} 0 \\ 0 \end{bmatrix},$$

is indicated in Table 6.4. From $n = 2$, the policy can be seen to have converged to $[A, B]$. For this policy, application of (6.19) results in

$$y = \begin{bmatrix} 1.176 \\ 1.59 \end{bmatrix}.$$

**Table 6.4**

| $n$ | $y_1$ | $y_2$ | $0.1y_1+0.1y_2+0.9$ | $0.4y_1+0.3y_2+0.6$ | $\hat{U}_2$ | $0.5y_1+0.2y_2+0.7$ | $0.1y_1+0.8y_2+0.2$ | $\hat{U}_1$ |
|---|---|---|---|---|---|---|---|---|
| 0 | 0 | 0 | 0.9 | **0.6** | *B* | 0.7 | **0.2** | *B* |
| 1 | 0.6 | 0.2 | 0.980 | **0.900** | *B* | 1.040 | **0.420** | *B* |
| 2 | 0.900 | 0.420 | **1.032** | 1.086 | *A* | 1.234 | **0.626** | *B* |
| 3 | 1.032 | 0.626 | **1.066** | 1.199 | *A* | 1.341 | **0.803** | *B* |
| 4 | 1.066 | 0.803 | **1.087** | 1.266 | *A* | 1.393 | **0.943** | *B* |
| 5 | 1.087 | 0.943 | **1.103** | 1.316 | *A* | 1.431 | **1.063** | *B* |
| 6 | 1.103 | 1.063 | **1.117** | 1.360 | *A* | 1.451 | **1.161** | *B* |
| 7 | 1.117 | 1.161 | **1.128** | 1.395 | *A* | 1.490 | **1.241** | *B* |
| 8 | 1.128 | 1.241 | **1.137** | 1.424 | *A* | 1.512 | **1.306** | *B* |
| 9 | 1.137 | 1.306 | **1.447** | 1.447 | *A* | 1.530 | **1.360** | *B* |
| 10 | 1.447 | 1.360 | | | | | | |

That this is indeed the optimal solution can be verified by substitution in (6.20), to obtain

$$1.176 = \min[0.1 \times 1.176 + 0.1 \times 1.59 + 0.9 = 1.176,$$
$$0.4 \times 1.176 + 0.3 \times 1.59 + 0.6 = 1.547],$$

$$1.588 = \min[0.5 \times 1.176 + 0.2 \times 1.59 + 0.7 = 1.606,$$
$$0.1 \times 1.176 + 0.8 \times 1.59 + 0.2 = 1.588].$$

In addition, this is the result found directly by continuing in Table 6.3.

**Remark.** As before, convergence of this iterative process can be accelerated by approximation in policy space. If an iteration in policy space, analogous to that in Section 3.4.4, is used here, convergence is assured simply by choosing an admissible policy to initialize the process.

## 6.4 Discrete Stochastic Processes with Discrete States Which Are Not Completely Measurable

In this last section we shall consider processes in which the state is not completely measurable, in the particular case with states that are discrete. Even though the solution which we shall obtain is of limited practical use, we shall see that on the theoretical level it is sufficient to deal with the complexities of the general case treated in Section 5.5.

### *6.4.1 Definition of the Problem*

We shall consider a system analogous to that of the preceding sections, in which the state at time $n$, $e_n$, can be represented as an element $i$ of a discrete set, $i \in \{1, \ldots, M\}$. The evolution of the states in time is governed by $P_{ij}(\boldsymbol{u}, n)$, the probabilities that, if the state at time $n$ is $i$, and if the control $\boldsymbol{u}$ is applied, the resulting state at time $n+1$ will be $j$.

In addition, the state $i$ is not directly measurable. Only the value of some output function $s_n$ is known, which at time $n$ takes on discrete values $m$, depending stochastically on the state through the probabilities

$$\tilde{\omega}_{mi}(n) = \text{prob}\{s_n = m \quad \text{if} \quad e_n = i\}. \tag{6.24}$$

Finally, there is defined an elementary return $r_{ij}(\boldsymbol{u}, n)$, which is the return resulting if transition from state $i$ to state $j$ occurs under the influence of control $\boldsymbol{u}$:

$$r_{ij}(\boldsymbol{u}, n) = \text{return relative to control } \boldsymbol{u} \qquad \text{when} \quad e_n = i \quad \text{becomes} \quad e_{n+1} = j. \tag{6.25}$$

An optimal control is to be found relative to time $n$, given only the values $s_0, \ldots, s_n$, such that the expected future return, from time $n$ until time $N$, is optimum.

We shall consider here a problem with bounded horizon and additive cost criterion, but the approach used can easily be applied to the problems with random horizon or terminal cost or both of these.

### 6.4.2 *Recurrence Relation of the a Priori Probabilities*

Let $\mathbf{I}_n$ denote the total information available about the process at time $n$, the sequence of applied controls and the measured outputs between times 0 and $n$. Let the *a priori* probability of being in state $j$, for a given $\mathbf{I}_n$, be $v_j(n)$:

$$v_j(n) = \text{prob}\{e_n = j \mid \mathbf{I}_n\}. \tag{6.26}$$

This probability summarizes all that is known about the state of the process at time $n$. We shall show that, starting from $v_i(0)$, which summarizes the initial information about the process, the probabilities $v_j(n)$ can be calculated recursively as a function of the successive values of $\boldsymbol{u}_n$ and $s_n$.

Applying the rule for combination of probabilities,

$$\text{prob}\{e_n = j \quad \text{and} \quad s_n = m \mid \mathbf{I}_{n-1}\} = \text{prob}\{e_n = j \mid s_n = m, \boldsymbol{u}, \mathbf{I}_{n-1}\} \times \text{prob}\{s_n = m \mid \mathbf{I}_{n-1}\}.$$

Further,

$$\begin{aligned}\text{prob}\{e_n = j \quad \text{and} \quad s_n = m \mid \mathbf{I}_{n-1}\} &= \sum_i \text{prob}\{e_{n-1} = i \mid \mathbf{I}_{n-1}\} \times \text{prob}\{e_n = j \mid e_{n-1} = i, \boldsymbol{u}\} \\ &\quad \times \text{prob}\{s_n = m \mid e_n = j\} \\ &= \sum_i v_i(n-1) P_{ij}(\boldsymbol{u}, n-1) \tilde{\omega}_{mi}(\boldsymbol{u}, n)\end{aligned} \tag{6.27}$$

$$\begin{aligned}\text{prob}\{s_n = m \mid \mathbf{I}_{n-1}\} &= \sum_i \sum_j v_i(n-1) P_{ij}(\boldsymbol{u}, n-1) \tilde{\omega}_{mi}(\boldsymbol{u}, n) \\ &= \rho_m(\boldsymbol{v}(n-1)).\end{aligned} \tag{6.28}$$

Relations (6.26), (6.27), and (6.28) yield

$$v_j(n) = \frac{\sum_i v_i(n-1) P_{ij}(\boldsymbol{u}, n-1) \tilde{\omega}_{mi}(\boldsymbol{u}, n)}{\sum_i \sum_j v_i(n-1) P_{ij}(\boldsymbol{u}, n-1) \tilde{\omega}_{mi}(\boldsymbol{u}, n)}. \tag{6.29}$$

It can be verified that these probabilities indeed sum to unity:

$$\sum_m \rho_m(\boldsymbol{v}) = 1,$$

$$\sum_j v_j(n) = 1.$$

Relation (6.29) will be written in condensed form as

$$\boldsymbol{v}_{n-1} \xrightarrow{m,\, \boldsymbol{u}_{n-1}} \boldsymbol{v}_n : \boldsymbol{v}(n) = \mathscr{T}_m[\boldsymbol{v}(n-1), \boldsymbol{u}_{n-1}]. \tag{6.30}$$

### 6.4.3 *The Optimal Recurrence Equation*

The optimal expected return is a function only of the time $n$, and the probability vector $\boldsymbol{v}(n)$, since the latter summarizes all information collected about the process from the initial instant.

Let $\hat{R}(\boldsymbol{v}, n)$ be the mathematical expectation of the total return from time $n$ up till time $N$, if the system is in state $i$ with probability $v_i$, taking into account the prior history of controls and outputs. The principle of optimality then leads to

$$\hat{R}(\boldsymbol{v}, n) = \underset{\boldsymbol{u} \in \Omega}{\text{opt}}\ \text{E[immediate return + future return]}. \tag{6.31}$$

The two parts of (6.31) can be evaluated as

$$\text{E[immediate return]} = \sum_i \left( \sum_j r_{ij}(\boldsymbol{u}, n) P_{ij}(\boldsymbol{u}, n) \right) v_i = \gamma(\boldsymbol{v}, \boldsymbol{u}), \tag{6.32}$$

$$\begin{aligned} \text{E[future return]} &= \sum_m \text{prob}\{s_{n+1} = m \mid \boldsymbol{u}\} \\ &\qquad \times \hat{R}[\boldsymbol{v}(n+1) \text{ for } \boldsymbol{v}(n) = \boldsymbol{v} \text{ and } s_{n+1} = m, n+1] \\ &= \sum_m \rho_m(\boldsymbol{v}) \hat{R}[\mathscr{T}_m(\boldsymbol{v}, \boldsymbol{u}), n+1], \end{aligned} \tag{6.33}$$

where $\rho_m(\boldsymbol{v})$ is as defined in (6.28). In the second relation, it is assumed that the future control policy is optimal, so that the future return is the optimal return.

Using (6.32), (6.33) in (6.31) yields finally

$$\hat{R}(\boldsymbol{v}, n) = \underset{\boldsymbol{u} \in \Omega}{\text{opt}} \{\gamma(\boldsymbol{v}, \boldsymbol{u}) + \sum_m \rho_m(\boldsymbol{v}) \hat{R}[\mathscr{T}_m(\boldsymbol{v}, \boldsymbol{u}), n+1]\} \tag{6.34}$$

which is a recurrence relation among the functions $\hat{R}(\boldsymbol{v}, n)$, to be initialized with

$$\hat{R}(\boldsymbol{v}, N) = 0. \tag{6.35}$$

The similarity of (6.34) to Eq. (6.18) of Section 5.2.3 might be noted, with (6.30) appearing as the equation of evolution of a process with $\boldsymbol{v}$ as state vector.

### 6.4.4 *Interpretation of the Results*

Equation (6.34) involves a function having the same number of arguments as the process in question has discrete states. In view of the nonlinear form of (6.30) only numerical methods are available for carrying out the solution, and, as a result, the barrier of dimensionality limits application of the procedure to processes with only four or five possible states.

The controller will use the recurrence relation (6.29), initialized with $\boldsymbol{v}_0$, and the control functions $\hat{g}_n(\boldsymbol{v})$ obtained by solving (6.34). The controller scheme is diagrammed in Fig. 6.4. Two elements are evident, one

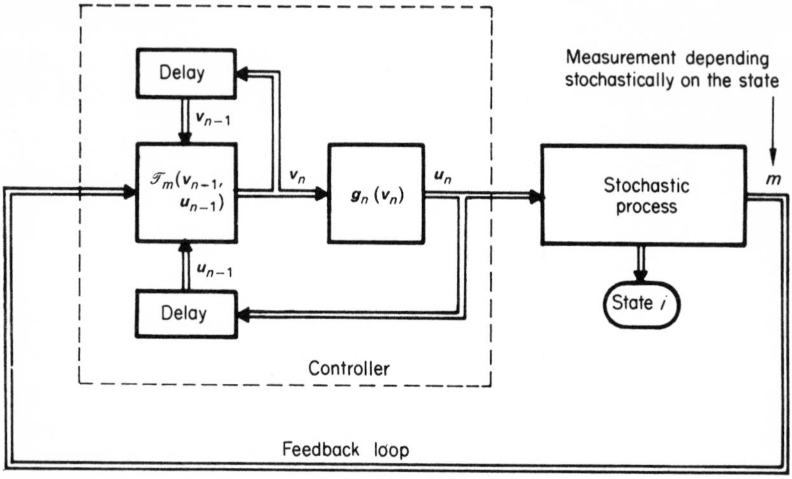


FIG. 6.4. Structure of the optimal controller.

filtering the information about the process, and the other forming the control signal. An analogous structure will be found later when linear systems with quadratic cost are considered.

From the theoretical point of view, the approach used here is quite interesting. By using the vector $\boldsymbol{v}$, we have been able to express the optimal controller as a function with a fixed number of arguments. This is a result arising from relation (6.30), which uses the measurements $s_n$ to generate the $\boldsymbol{v}_n$, which constitute a complete statistic.

In the case that the states are continuous, rather than discrete, the vector $\boldsymbol{v}$ passes from a vector space into a function space, and transforms according to a function $\mathrm{d}P(x)$, with

$$P(\boldsymbol{x}) = \text{prob}\{\text{state} \leq \boldsymbol{x} \,|\, \mathbf{I}_n\}. \tag{6.36}$$

This *a priori* probability distribution obeys a functional relation analogous to (6.29), with the summations being replaced by integrals. In this case, the optimal expected return can be expressed as a functional of $P(\mathbf{x})$, and even if a relation analogous to (6.34) can be written formally, it brings into play some functional relations, and direct practical use of the relation is impossible. In order to be able to have only functions to deal with, rather than functionals, the optimal return is expressed as a function of the available information, which reverts to the approach used in Chapter 5.

# PART 3 Numerical Synthesis of the Optimal Controller for a Linear Process

# Chapter 7 | General Discussion of the Problem

## 7.1 Definition of the Problem

We shall consider the general control problem diagrammed in Fig. 7.1. This system consists of a process, subject to control signals $\boldsymbol{u}(n)$ and some random perturbations, and producing outputs $\boldsymbol{s}(n)$; an input generator, producing a signal $\boldsymbol{e}(n)$; a perturbation generator, which acts on the process as well as altering the generated signals $\boldsymbol{e}(n)$ and $\boldsymbol{s}(n)$; and a controller, which constructs the control signals $\boldsymbol{u}(n)$ on the basis of the noisy signals $\tilde{\boldsymbol{e}}(n)$ and $\tilde{\boldsymbol{s}}(n)$.

The problem is to determine a controller which will result in optimal performance of the controlled system for given mathematical models of the process, input signal generator, and perturbations. Optimal performance is

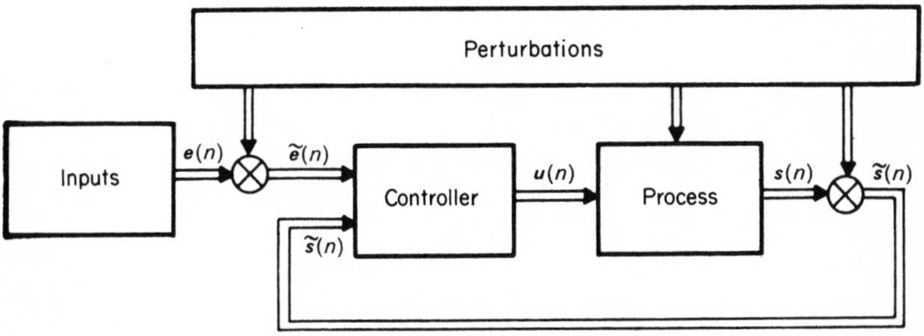


FIG. 7.1. General structure of the system considered.

characterized in terms of successful completion of some mission assigned in advance.

In this part of the book, we shall only consider the case in which performance is measured by a quadratic form, and the various mathematical models used are linear.

## 7.2 Mathematical Models of the Units

We shall first make precise the forms of the models involved in each part of the system. We shall then group these components into a single model of the entire system. To shorten the notation, we shall again drop the time index $n$, and write simply $Q$ for $Q(n)$, $Q^+$ for $Q(n+1)$, and $Q^-$ for $Q(n-1)$.

### *7.2.1 Model of the Process To Be Controlled*

With the above notational conventions, the mathematical model of the linear process to be controlled is a vector state equation of the form

$$\boldsymbol{x}^+ = \mathbf{A}\boldsymbol{x} + \mathbf{B}\boldsymbol{u} + \boldsymbol{p}_1, \tag{7.1}$$

$$\boldsymbol{s} = \mathbf{C}\boldsymbol{x}, \tag{7.2}$$

where $\boldsymbol{x}$ is the state vector, $\boldsymbol{u}$ the control vector, $\boldsymbol{s}$ the output vector, and $\boldsymbol{p}_1$ a perturbation vector. This system may be stationary or not, depending on whether the matrices $\mathbf{A}(n)$, $\mathbf{B}(n)$, $\mathbf{C}(n)$, denoted in all cases simply by $\mathbf{A}$, $\mathbf{B}$, $\mathbf{C}$, are or are not in fact constant.

A continuous-time linear process leads to a model of the discrete type (7.1), (7.2), if the control is stepped.

### *7.2.2 Model of the Input Generator*

The input signal can be either deterministic or random. In the deterministic case, if the class of input signals is linear, $\boldsymbol{e}(n)$ is the solution of a recurrence relation, without forcing term, which can be put in state vector form:

$$\boldsymbol{y}^+ = \mathbf{D}\boldsymbol{y}, \tag{7.3}$$

$$\boldsymbol{e} = \mathbf{D}'\boldsymbol{y}, \tag{7.4}$$

where $\boldsymbol{y}$ is the state vector and $\boldsymbol{e}$ the output.

In the random case, because of the linear nature of the problem, $\boldsymbol{e}$ can be generated by a Markov process of the type

$$\boldsymbol{y}^{+} = \mathbf{D}\boldsymbol{y} + \boldsymbol{\beta}, \tag{7.5}$$

$$\boldsymbol{e} = \mathbf{D}'\boldsymbol{y}, \tag{7.6}$$

where $\boldsymbol{\beta}$ is a pseudowhite noise, uncorrelated in time, having zero mean and correlation matrix

$$\mathrm{E}[\boldsymbol{\beta}\boldsymbol{\beta}^{\mathrm{T}}] = \boldsymbol{\Phi}_{\beta}. \tag{7.7}$$

The random and deterministic cases can be grouped together, since the latter corresponds to the particular choice $\boldsymbol{\Phi}_{\beta} = \mathbf{0}$. A random vector does not appear in (7.6), since the presence of a "white" part in the signal $\boldsymbol{e}$ would not correspond to a realistic problem.

### *7.2.3 Perturbation Generator*

The perturbations $\boldsymbol{p}_1$, $\boldsymbol{p}_2$, and $\boldsymbol{p}_3$, being, respectively, the random disturbance acting on the process, and the errors in measurement of $\boldsymbol{s}$ and $\boldsymbol{e}$, will be taken to be random, and generated by linear Markov processes of the type

$$\boldsymbol{z}^{+} = \mathbf{M}\boldsymbol{z} + \boldsymbol{\delta}', \tag{7.8}$$

$$\boldsymbol{p}_i = \mathbf{K}_i\boldsymbol{z} + \mathbf{K}_i'\boldsymbol{\delta}, \tag{7.9}$$

where $\boldsymbol{z}$ is the state vector of the Markov process, $\boldsymbol{p}_i$, $i = 1, 2, 3$, are the perturbations, and $\boldsymbol{\delta}$ and $\boldsymbol{\delta}'$ are pseudowhite, independent, noise signals, uncorrelated in time, with zero means and correlation matrices $\boldsymbol{\Phi}_{\delta}$ and $\boldsymbol{\Phi}_{\delta'}$. The use of a common dynamic model (7.8) allows consideration of the case in which $\boldsymbol{p}_1$, $\boldsymbol{p}_2$, and $\boldsymbol{p}_3$ are dependent.

## 7.3 The Canonical Model

The preceding models have analogous structures, and can be grouped so as to provide a model of the complete process. The result is the scheme shown in Fig. 7.2, and the following recursion relations:

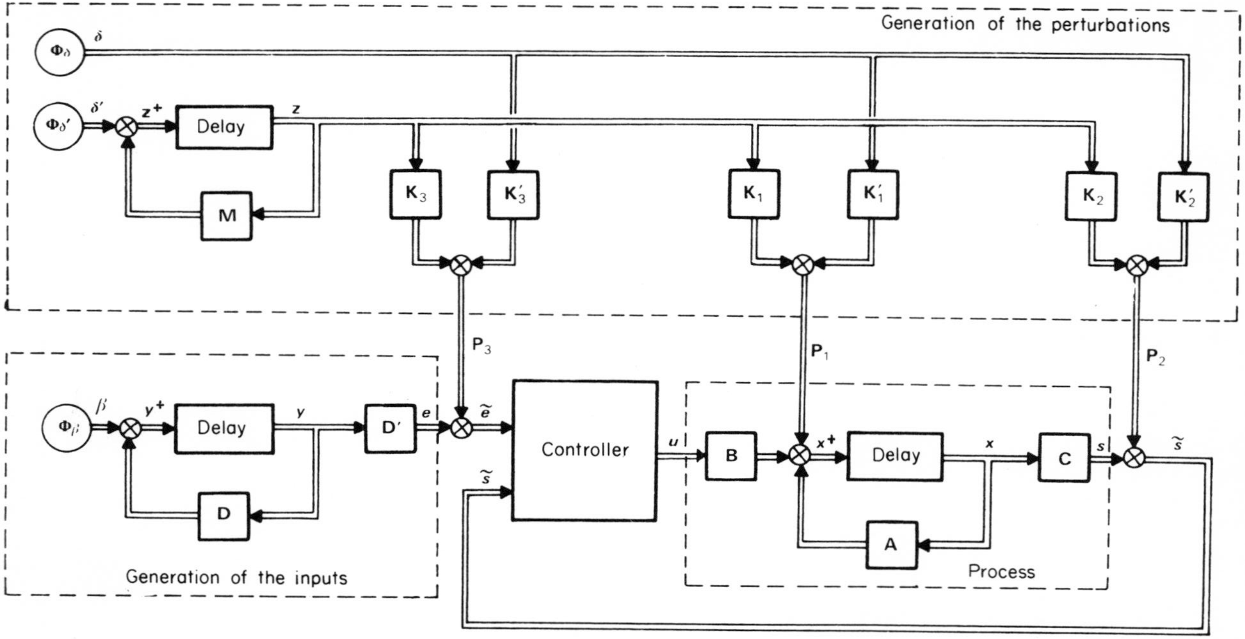


FIG. 7.2. Structure of the models of the system elements.

$$\begin{bmatrix} x \\ y \\ z \end{bmatrix}^+ = \begin{bmatrix} \mathbf{A} & \mathbf{0} & \mathbf{K}_1 \\ \mathbf{0} & \mathbf{D} & \mathbf{0} \\ \mathbf{0} & \mathbf{0} & \mathbf{M} \end{bmatrix} \begin{bmatrix} x \\ y \\ z \end{bmatrix} + \begin{bmatrix} \mathbf{B} \\ \mathbf{0} \\ \mathbf{0} \end{bmatrix} u + \begin{bmatrix} \mathbf{0} & \mathbf{0} \\ \mathbf{1} & \mathbf{0} \\ \mathbf{0} & \mathbf{1} \end{bmatrix} \begin{bmatrix} \beta \\ \delta' \end{bmatrix} + \begin{bmatrix} \mathbf{K}_1' \\ \mathbf{0} \\ \mathbf{0} \end{bmatrix} \delta \quad (7.10)$$

$$\begin{bmatrix} \tilde{e} \\ \tilde{s} \end{bmatrix} = \begin{bmatrix} \mathbf{0} & \mathbf{D}' & \mathbf{K}_3 \\ \mathbf{C} & \mathbf{0} & \mathbf{K}_2 \end{bmatrix} \begin{bmatrix} x \\ y \\ z \end{bmatrix} + \begin{bmatrix} \mathbf{K}_3' \\ \mathbf{K}_2' \end{bmatrix} \delta . \quad (7.11)$$

Letting

$$X = \begin{bmatrix} x \\ y \\ z \end{bmatrix}, \qquad I = \begin{bmatrix} \tilde{e} \\ \tilde{s} \end{bmatrix}, \qquad \Delta = \begin{bmatrix} \beta \\ \delta' \\ \delta \end{bmatrix},$$

$$\mathbf{F} = \begin{bmatrix} \mathbf{A} & \mathbf{0} & \mathbf{K}_1 \\ \mathbf{0} & \mathbf{D} & \mathbf{0} \\ \mathbf{0} & \mathbf{0} & \mathbf{M} \end{bmatrix}, \qquad \mathbf{H} = \begin{bmatrix} \mathbf{B} \\ \mathbf{0} \\ \mathbf{0} \end{bmatrix}, \qquad \mathbf{G} = \begin{bmatrix} \mathbf{0} & \mathbf{D}' & \mathbf{K}_3 \\ \mathbf{C} & \mathbf{0} & \mathbf{K}_2 \end{bmatrix},$$

$$\mathbf{\Pi} = \begin{bmatrix} \mathbf{0} & \mathbf{0} & \mathbf{K}' \\ \mathbf{1} & \mathbf{0} & \mathbf{0} \\ \mathbf{0} & \mathbf{1} & \mathbf{0} \end{bmatrix}, \qquad \mathbf{\Pi}' = \begin{bmatrix} \mathbf{0} & \mathbf{0} & \mathbf{K}_3' \\ \mathbf{0} & \mathbf{0} & \mathbf{K}_2' \end{bmatrix}, \qquad \mathbf{\Phi} = \begin{bmatrix} \mathbf{\Phi}_\beta & \mathbf{0} & \mathbf{0} \\ \mathbf{0} & \mathbf{\Phi}_{\delta'} & \mathbf{0} \\ \mathbf{0} & \mathbf{0} & \mathbf{\Phi}_\delta \end{bmatrix},$$

Eqs. (7.10) and (7.11) can be combined into the canonical model

$$X^+ = \mathbf{F}X + \mathbf{H}u + \mathbf{\Pi}\Delta, \quad (7.12)$$

$$I = \mathbf{G}X + \mathbf{\Pi}'\Delta, \quad (7.13)$$

where $X$ is the global state vector, $I$ the information vector, and $\Delta$ a pseudowhite noise vector with correlation matrix $\mathbf{\Phi}$. Thus if the various linearity conditions hold, any system with the general structure indicated in Fig. 7.1 can be put into the form (7.12), (7.13), which corresponds to a regulator problem.

On the practical level, use of a model such as (7.12), (7.13) passes over many steps, some of which can involve experimental or numerical difficulties. These are summarized in Fig. 7.3.

The first step is identification of the various elements of the model, based on the results of tests made on the actual process. Two approaches are the most common. The first method consists in assuming a structure for the model, based on theoretical considerations. To determine the values of the various coefficients in the model, a process of parameter identification is carried out, using methods such as described by Boudarel *et al.* [1, Vol. 2]. Sometimes a lack of consistency in the results leads to adjustment of the theoretical model in order better to account for the test results. The second

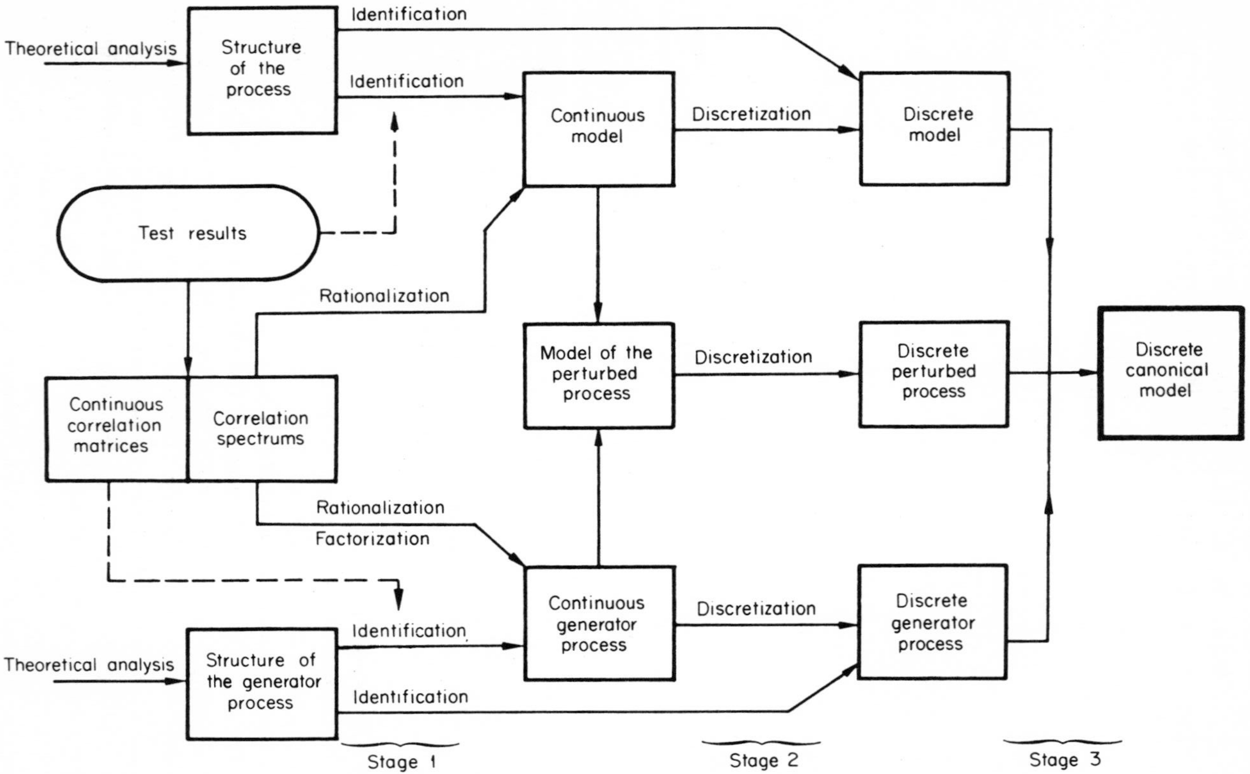


FIG. 7.3. Principal stages in computing the canonical system model.

general method is a stochastic identification process, which does not require setting up a preliminary model. However, in order to obtain a state-vector model, use must be made of transfer functions calculated as rational fractions, which constitutes an *a posteriori* characterization.

At the second stage, the continuous model is converted to a discrete model, using the methods of Boudarel *et al.* [1, Vol. 1, Chaps. 4 and 14]. If the continuous process to be controlled is subject to continuous perturbation signals at its input, it is necessary to discretize together the process and the system generating the perturbations. The final step is to group together the various partial models to obtain the complete canonical model.

## 7.4 The System Objective

The system starts in an initial state which is known only through $\tilde{e}$ and $\tilde{s}$. If no perturbations are present, conditions can be imposed on the terminal state, such as $\boldsymbol{x} = \boldsymbol{0}$. In the random case, however, such terminal conditions have no meaning.

The process evolution will be characterized by an elementary return which is a quadratic form in $\boldsymbol{x}, \boldsymbol{y}$, and $\boldsymbol{u}$. The total return is thus of the form

$$\mathscr{R} = \sum_{n=0}^{N-1} \begin{bmatrix} \boldsymbol{x}(n) \\ \boldsymbol{y}(n) \\ \boldsymbol{u}(n) \end{bmatrix}^{\mathrm{T}} \mathbf{R}(n) \begin{bmatrix} \boldsymbol{x}(n) \\ \boldsymbol{y}(n) \\ \boldsymbol{u}(n) \end{bmatrix}, \tag{7.14}$$

where $\mathbf{R}(n)$ is a symmetric nonnegative definite matrix.

In some problems, there can arise global quadratic constraints with a structure analogous to (7.14):

$$\mathscr{C}^i = \sum_{n=0}^{N-1} \begin{bmatrix} \boldsymbol{x}(n) \\ \boldsymbol{y}(n) \\ \boldsymbol{u}(n) \end{bmatrix}^{\mathrm{T}} \mathbf{P}^i(n) \begin{bmatrix} \boldsymbol{x}(n) \\ \boldsymbol{y}(n) \\ \boldsymbol{u}(n) \end{bmatrix} \leq \gamma_i . \tag{7.15}$$

The matrices in (7.14) and (7.15) depend on the requirements imposed on the controlled system. Using the vector $\boldsymbol{X}$ defined in the preceding section, (7.14) and (7.15) can be written

$$\mathscr{R} = \sum_{n=0}^{N-1} \begin{bmatrix} \boldsymbol{X} \\ \boldsymbol{u} \end{bmatrix}^{\mathrm{T}} \mathbf{R} \begin{bmatrix} \boldsymbol{X} \\ \boldsymbol{u} \end{bmatrix} = \sum_{n=0}^{N-1} [\boldsymbol{X}^{\mathrm{T}}\mathbf{R}_1\boldsymbol{X} + 2\boldsymbol{X}^{\mathrm{T}}\mathbf{R}_2\boldsymbol{u} + \boldsymbol{u}^{\mathrm{T}}\mathbf{R}_3\boldsymbol{u}], \tag{7.16}$$

$$\mathscr{C}^i = \sum_{n=0}^{N-1} \boldsymbol{X}^{\mathrm{T}}\mathbf{P}_1{}^i\boldsymbol{X} + 2\boldsymbol{X}^{\mathrm{T}}\mathbf{P}_2{}^i\boldsymbol{u} + \boldsymbol{u}^{\mathrm{T}}\mathbf{P}_3{}^i\boldsymbol{u}. \tag{7.17}$$

## 7.5 Problem Types

Within the limits of the hypotheses of linearity of the processes and quadratic cost, all control problems can be put in the form of a process (7.12), (7.13), with criterion (7.16) and constraints (7.17). Such unification, however, is only formal since, even if all problems are fused together into a single general problem, regulation in the presence of pseudowhite perturbations with the state not completely measurable, each particular class of problems still retains its own properties and difficulties.

In order to treat the general problem we shall proceed in stages of increasing difficulty and treat successively the following specific problems:

a. The deterministic regulator problem, with completely measurable state. This corresponds to $\mathbf{G} = \mathbf{1}$, $\mathbf{\Pi} = \mathbf{0}$, $\mathbf{\Pi}' = \mathbf{0}$.

b. The stochastic regulator problem, with measurable state. This corresponds to $\mathbf{G} = \mathbf{1}$, $\mathbf{\Pi} \neq \mathbf{0}$, $\mathbf{\Pi}' \neq \mathbf{0}$.

c. The general problem, corresponding to $\mathbf{G} \neq \mathbf{1}$, $\mathbf{\Pi} \neq \mathbf{0}$, and $\mathbf{\Pi}' \neq \mathbf{0}$.

Problem a will be the subject of Chapter 8, and in Chapter 9 we shall see how the solution to the deterministic problem can be extended to the stochastic case (problems b and c).

**Remark.** Suppose that the cost criterion is to be insensitive to variations of some parameter in the mathematical model of the process. As discussed in Section 2.8, this requires introduction of additional state and control variables satisfying a linear recurrence system, and a constraint corresponding to the condition that the sensitivity in question be zero. Since the cost criterion is quadratic and the process linear, this constraint will also be quadratic. Thus, at the price of increasing the order of the system and introducing a quadratic constraint, the canonical structure used above can be retained for problems with a sensitivity condition. We shall thus not treat this type of problem explicitly, since it lies within the framework of the general problem treated.

# Chapter 8 Numerical Optimal Control of a Measurable Deterministic Process

In this chapter, we shall consider the simplest class of problems of interest, those in which the process is described by

$$x_{n+1} = \mathbf{F}_n x_n + \mathbf{H}_n u_n \tag{8.1}$$

and in which the cost criterion is quadratic:

$$\mathscr{R}_m{}^N = \sum_{n=m}^{N-1} [x_n{}^{\mathrm{T}}\mathbf{R}_1{}^n x_n + 2x_n{}^{\mathrm{T}}\mathbf{R}_2{}^n u_n + u_n{}^{\mathrm{T}}\mathbf{R}_3{}^n u_n]. \tag{8.2}$$

It was established in Section 4.6 that in this case the optimal control is linear,

$$\hat{u}_n = -\mathbf{L}_n x_n, \tag{8.3}$$

and that the attained optimal cost is quadratic:

$$\mathscr{R}_m{}^N = x_m{}^{\mathrm{T}}\mathbf{Q}_m x_m. \tag{8.4}$$

These relations hold true only if the terminal state is free, or an equilibrium state. If that is not the case, additional terms are needed [see (4.36), (4.40), and (4.41)]. The matrices $\mathbf{L}$ and $\mathbf{Q}$ satisfy the system of recursion relations

$$\mathbf{L} = [\mathbf{R}_3 + \mathbf{H}^{\mathrm{T}}\mathbf{Q}^{-}\mathbf{H}]^{-1}[\mathbf{R}_2{}^{\mathrm{T}} + \mathbf{H}^{\mathrm{T}}\mathbf{Q}^{-}\mathbf{F}], \tag{8.5}$$

$$\mathbf{Q} = \mathbf{R}_1 + \mathbf{F}^{\mathrm{T}}\mathbf{Q}^{-}\mathbf{F} - [\mathbf{R}_2 + \mathbf{F}^{\mathrm{T}}\mathbf{Q}^{-}\mathbf{H}]\mathbf{L}. \tag{8.6}$$

In these equations, time is counted backward from the final time $N$. When the terminal state is free, (8.5), (8.6) should be initialized with

$$\mathbf{Q}_0 = \mathbf{0}. \tag{8.7}$$

A flow chart of the controller corresponding to (8.3) is shown in Fig. 8.1. Note the position of the input–output operations with respect to the actual calculations.

When the terminal state $\boldsymbol{x}_{\mathrm{f}}$ is required to satisfy certain independent linear conditions,

$$\boldsymbol{\Gamma x}_{\mathrm{f}} = \boldsymbol{a}, \tag{8.8}$$

it is first necessary to solve the problem of terminal regulation. This question was touched upon in Section 4.6, but we shall treat it in greater detail in the following section.

In the second section we shall examine the problem of regulation in minimum time, with bounded total energy, and we shall show that the solution of this problem is related to that of the first problem.

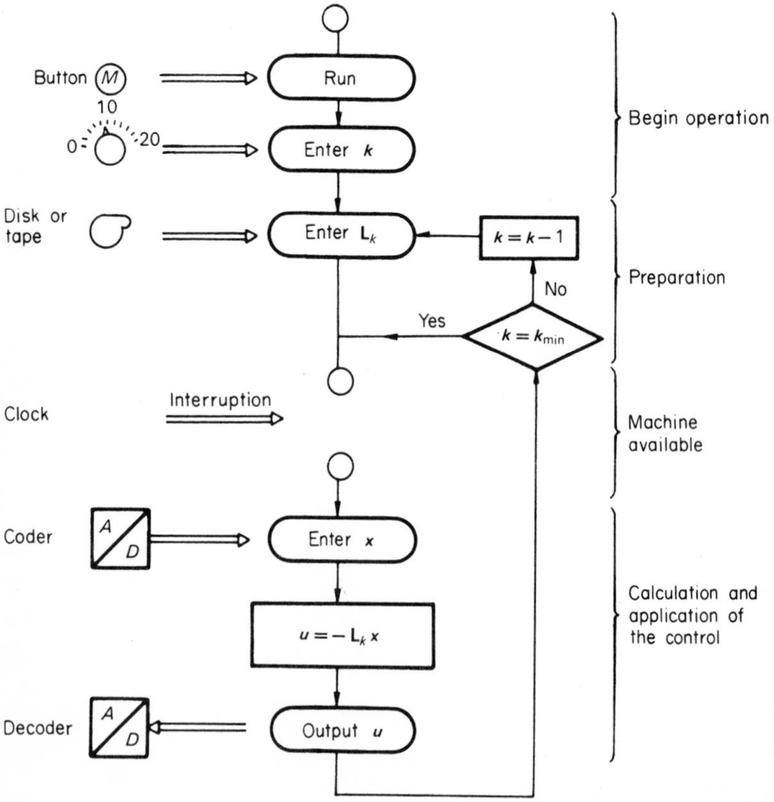


FIG. 8.1. Flow chart of the linear optimal controller.

In the third section quadratic constraints of the form (8.2) are introduced, and the solution is found using Lagrange multipliers.

The case of an infinite horizon is examined briefly in the last section of the chapter.

## 8.1 The Effect of Terminal Constraints

For simplicity, we shall assume that the process is stationary. However, the method can be extended easily to the more general case. Let $p$ be the dimension of the state vector, $\boldsymbol{x} \in \mathbf{R}^p$, let $q < p$ be the number of terminal constraints, $\boldsymbol{a} \in \mathbf{R}^q$, and let $l$ be the dimension of the control vector, $\boldsymbol{u} \in \mathbf{R}^l$.

### *8.1.1 The General Case*

Let $\boldsymbol{X}$ be the value of the state at time $N - k$. Counting time in the reverse direction, and using (8.1), the terminal condition (8.8) can be written

$$\boldsymbol{a} = \boldsymbol{\Gamma}(\mathbf{F}^k\mathbf{X} + \sum_{n=1}^{k} \mathbf{H}\mathbf{F}^{i-1}\boldsymbol{u}_i),$$

or

$$\boldsymbol{a} - \mathbf{N}_k \boldsymbol{X} = \mathbf{M}_k \boldsymbol{U}_k, \tag{8.9}$$

where

$$\mathbf{N}_k = \boldsymbol{\Gamma}\mathbf{F}^k, \tag{8.10}$$

$$\mathbf{M}_k = \boldsymbol{\Gamma}[\mathbf{H}, \mathbf{HF}, \ldots, \mathbf{HF}^{k-1}], \tag{8.11}$$

$$\boldsymbol{U}_k = \begin{bmatrix} \boldsymbol{u}_1 \\ \boldsymbol{u}_2 \\ \vdots \\ \boldsymbol{u}_k \end{bmatrix}. \tag{8.12}$$

Note that, on account of the time reversal used, the first control to be applied should be $\boldsymbol{u}_k$, and the last should be $\boldsymbol{u}_1$.

Depending on the rank of $\mathbf{M}_k$, as compared to $kl$, the linear system (8.9) has no solution, a unique solution, or many solutions. If the system is controllable (see Boudarel *et al.* [1, Vol. 1, Ch. 6]), there exists a $k_0$ such that

$$\operatorname{rank}[\mathbf{H}, \mathbf{HF}, \ldots, \mathbf{HF}^{k_0-1}] = p. \tag{8.13}$$

Then

$$\operatorname{rank}[\mathbf{M}_{k_0}] = \min(p, q) = q\,, \tag{8.14}$$

and for all $k$ such that

$$k \geq k_0 \qquad \text{and} \qquad kl \geq q\,, \tag{8.15}$$

the system (8.9) has infinitely many solutions from among which we wish to choose the one that minimizes the criterion (8.2).

The quadratic structure of the cost criterion requires that, for a given control law, the total return also be quadratic in $\boldsymbol{X}$ and $\boldsymbol{U}_k$:

$$\mathscr{R}_k(\boldsymbol{X}, \boldsymbol{U}_k) = \boldsymbol{X}^{\mathrm{T}}\mathbf{D}_k\boldsymbol{X} + 2\boldsymbol{X}^{\mathrm{T}}\boldsymbol{\Omega}_k\boldsymbol{U}_k + \boldsymbol{U}_k{}^{\mathrm{T}}\boldsymbol{\Lambda}_k\boldsymbol{U}_k\,, \tag{8.16}$$

where $\mathbf{D}_k$ is a matrix of dimensions $p \times p$, $\boldsymbol{\Omega}_k$ is $p \times kl$, and $\boldsymbol{\Lambda}_k$ is $kl \times kl$. These matrices can be evaluated by expanding

$$\mathscr{R}_{k+1}(\boldsymbol{X}, \boldsymbol{U}_{k+1}) = \mathscr{R}_k(\mathbf{F}\boldsymbol{X} + \mathbf{H}\boldsymbol{u}_{k+1}, \boldsymbol{U}_k)\,. \tag{8.17}$$

The result is the recursive relations

$$\mathbf{D}^{+} = \mathbf{R}_1 + \mathbf{F}^{\mathrm{T}}\mathbf{D}\mathbf{F}\,, \tag{8.18}$$

$$\boldsymbol{\Omega}^{+} = [\mathbf{F}\boldsymbol{\Omega}, \mathbf{F}^{\mathrm{T}}\mathbf{D}\mathbf{H} + \mathbf{R}_2]\,, \tag{8.19}$$

$$\boldsymbol{\Lambda}^{+} = \begin{bmatrix} \boldsymbol{\Lambda}, & \boldsymbol{\Omega}^{\mathrm{T}}\mathbf{H} \\ \mathbf{H}^{\mathrm{T}}\boldsymbol{\Omega}, & \mathbf{R}_3 \end{bmatrix}, \tag{8.20}$$

which are to be initialized with

$$\mathbf{D} = \mathbf{R}_1, \qquad \boldsymbol{\Omega} = \mathbf{R}_2\,, \qquad \boldsymbol{\Lambda} = \mathbf{R}_3\,. \tag{8.21}$$

If there exists a $k$ such that

$$kl = q \qquad \text{and} \qquad |\mathbf{M}_k| \neq 0\,,$$

the system (8.9) then has a unique solution $\boldsymbol{U}_k$, which can be substituted into (8.16) to obtain $\mathbf{Q}_k$. In Section 8.1.2 we shall present some examples in which there exists a $k$ beyond which the regulator problem completely determines the control law.

In general, the smallest $k$ for which (8.9) has solutions is found, and then $\boldsymbol{U}_k$ is found to minimize (8.16), considering (8.9) as an equality constraint on the minimization problem. This nonlinear programming problem can be solved in literal form with the use of Lagrange multipliers.

Introducing the modified criterion

$$\varphi(\boldsymbol{U}_k) = \boldsymbol{X}^{\mathrm{T}}\mathbf{D}_k\boldsymbol{X} + 2\boldsymbol{X}^{\mathrm{T}}\boldsymbol{\Omega}_k\boldsymbol{U}_k + \boldsymbol{U}_k{}^{\mathrm{T}}\boldsymbol{\Lambda}_k\boldsymbol{U}_k + \boldsymbol{\lambda}^{\mathrm{T}}(\boldsymbol{a} - \mathbf{N}_k\boldsymbol{X} - \mathbf{M}_k\boldsymbol{U}_k)\,, \tag{8.22}$$

and applying the first-order optimality condition, leads to

$$\varphi_U = 2\boldsymbol{\Omega}_k{}^{\mathrm{T}}X + 2\boldsymbol{\Lambda}_k \hat{U}_k - \mathbf{M}_k{}^{\mathrm{T}}\boldsymbol{\lambda} = \boldsymbol{0},$$

or

$$\hat{U}_k = \boldsymbol{\Lambda}_k^{-1}[\tfrac{1}{2}\mathbf{M}_k{}^{\mathrm{T}}\boldsymbol{\lambda} - \boldsymbol{\Omega}_k{}^{\mathrm{T}}X].$$

To determine $\boldsymbol{\lambda}$, $\hat{U}_k$ is substituted into (8.9) to yield

$$\boldsymbol{a} - \mathbf{N}_k X = \mathbf{M}_k \boldsymbol{\Lambda}_k^{-1}[\tfrac{1}{2}\mathbf{M}_k{}^{\mathrm{T}}\boldsymbol{\lambda} - \boldsymbol{\Omega}_k{}^{\mathrm{T}}X],$$

from which

$$\boldsymbol{\lambda} = 2[\mathbf{M}_k \boldsymbol{\Lambda}_k^{-1}\mathbf{M}_k{}^{\mathrm{T}}]^{-1}[\boldsymbol{a} + (\mathbf{M}_k \boldsymbol{\Lambda}_k^{-1}\boldsymbol{\Omega}_k{}^{\mathrm{T}} - \mathbf{N}_k)X].$$

Thus, finally,

$$\begin{aligned}\hat{U}_k &= \boldsymbol{\Lambda}_k^{-1}\{\mathbf{M}_k{}^{\mathrm{T}}[\mathbf{M}_k \boldsymbol{\Lambda}_k^{-1}\mathbf{M}_k{}^{\mathrm{T}}]^{-1}[\boldsymbol{a} + (\mathbf{M}_k \boldsymbol{\Lambda}_k^{-1}\boldsymbol{\Omega}_k{}^{\mathrm{T}} - \mathbf{N}_k)X] - \boldsymbol{\Omega}_k{}^{\mathrm{T}}X\} \\ &= \mathbf{K}_k{}^1\boldsymbol{a} + \mathbf{K}_k{}^2X. \qquad (8.23)\end{aligned}$$

In order that $\hat{U}_k$ indeed minimize the criterion, it is necessary that the proper second-order condition be satisfied. The condition, in this case, is that

$$\begin{bmatrix}\mathbf{R}_1 & \mathbf{R}_2 \\ \mathbf{R}_2{}^{\mathrm{T}} & \mathbf{R}_3\end{bmatrix} \qquad (8.24)$$

be nonnegative definite.

In order that such an optimal solution exist and be unique, it is necessary that

$$|\boldsymbol{\Lambda}_k| \neq 0, \qquad (8.25)$$

$$|\mathbf{M}_k \boldsymbol{\Lambda}_k^{-1}\mathbf{M}_k{}^{\mathrm{T}}| \neq 0. \qquad (8.26)$$

Further, if (8.25) is satisfied, then the controllability of the system ensures (8.26). This is because, from (8.25),

$$\operatorname{rank}[\boldsymbol{\Lambda}_k] = kl,$$

and from (8.11)

$$\operatorname{rank}[\mathbf{M}_k] = q.$$

Then, on account of (8.15),

$$\operatorname{rank}[\mathbf{M}_k \boldsymbol{\Lambda}_k^{-1}\mathbf{M}_k{}^{\mathrm{T}}] = \min(q, lk) = q.$$

But this last matrix is square, and of order $q$. Hence it indeed has an inverse, and (8.26) holds true.

Interpretation of condition (8.25) is delicate, since not only **F** and **H**, but also $\mathbf{R}_1$, $\mathbf{R}_2$, and $\mathbf{R}_3$, are involved. In particular, it is not necessary that

$$|\mathbf{R}_3| \neq 0 .$$

If the problem posed is sensible, (8.15), (8.25), and (8.26) will be compatible, and there will exist a minimum value $\tilde{k}$ for which the regulation problem can be solved. Using this $\tilde{k}$, and substituting (8.24) into (8.16), yields the general form of the return. The return will be quadratic in $X$ only if $\boldsymbol{a}$ is zero, or can be made so by translation of the axes, which means that $\boldsymbol{a}$ belongs to an equilibrium subspace of the state space. If this is not the case, the return $\mathscr{R}$ has a constant term and a term linear in $X$, and the more general formulas (8.38) to (8.41) of Chapter 4 must be used.

Thus, for initialization of the optimal recurrence equations in the presence of terminal conditions, the scheme of Fig. 8.2 is appropriate. This summarizes the above steps. The matrix inversion algorithm (or the linear system solution algorithm) should give some indication if the determinant of the corresponding matrix is zero. (See [32] for material on the numerical aspects of these questions.)

### *8.1.2 Specific Problems*

a. Let us consider the case in which $\boldsymbol{a} = 0$ and $u$ is scalar ($l = 1$). Conditions (8.25) and (8.26) are then satisfied for

$$\tilde{k} = q . \tag{8.27}$$

If the final state is entirely specified,

$$q = p .$$

Finally, if the system is controllable, $\mathbf{M}_p$ has an inverse, and (8.23) simplifies considerably:

$$\hat{U}_p \equiv -\mathbf{M}_p^{-1}\mathbf{N}_p X = -[\mathbf{H}, \mathbf{FH}, \ldots, \mathbf{F}^{p-1}\mathbf{H}]^{-1}\mathbf{F}^p X . \tag{8.28}$$

This formula could also be written directly, since in the present case (8.9) has only a single solution.

b. Let us now consider the case in which $\boldsymbol{a} = \boldsymbol{0}$, and the criterion depends only on $\boldsymbol{u}$. Then

$$\mathbf{R}_1 = \mathbf{0}, \qquad \mathbf{R}_2 = \mathbf{0},$$

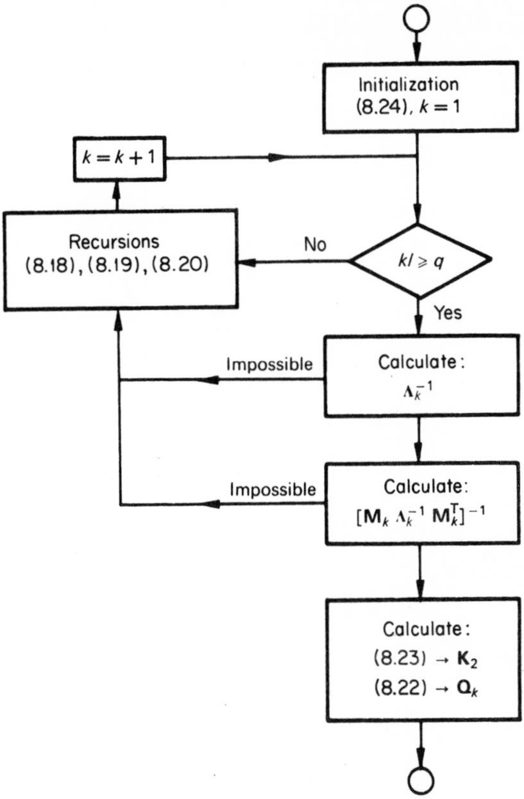


FIG. 8.2. Initialization procedure in the case of terminal constraints.

and

$$\mathbf{D}_k \equiv \mathbf{0}, \qquad \mathbf{\Omega}_k \equiv \mathbf{0},$$

$$\mathbf{\Lambda}_k = \begin{bmatrix} \mathbf{R}_3 & & \mathbf{0} \\ & \ddots & \\ \mathbf{0} & & \mathbf{R}_3 \end{bmatrix} \rightarrow \mathbf{\Lambda}_k^{-1} = \begin{bmatrix} \mathbf{R}_3^{-1} & & \mathbf{0} \\ & \ddots & \\ \mathbf{0} & & \mathbf{R}_3^{-1} \end{bmatrix}. \tag{8.29}$$

Condition (8.25) is satisfied if

$$|\mathbf{R}_3| \neq 0. \tag{8.30}$$

Since $\mathbf{R}_3$ should be nonnegative definite, in order that the second-order condition (8.24) hold true, the result of (8.30) is that $\mathbf{R}_3$ should be strictly positive definite. This implies that all components of $\boldsymbol{u}$ enter independently into the criterion.

In this case, (8.23) simplifies to

$$\hat{U}_k = -\Lambda_k^{-1}\mathbf{M}_k[\mathbf{M}_k\Lambda_k^{-1}\mathbf{M}_k^{\mathrm{T}}]^{-1}\mathbf{N}_k X, \tag{8.31}$$

which, when substituted into (8.16), results in

$$\mathbf{Q}_k = \mathbf{N}_k^{\mathrm{T}}[\mathbf{M}_k\Lambda_k^{-1}\mathbf{M}_k^{\mathrm{T}}]^{-1}\mathbf{N}_k. \tag{8.32}$$

In the particular case that the final state is entirely specified, and the system is controllable in $k = p/l$ steps, the matrix $\mathbf{M}_k$ is square and invertible. The system (8.9) then has a unique solution, and the result analogous to (8.28) is recovered, the corresponding matrix $\mathbf{Q}_k$ being

$$\mathbf{Q}_k = [\mathbf{M}_k^{-1}\mathbf{N}_k]^{\mathrm{T}}\Lambda_k[\mathbf{M}_k^{-1}\mathbf{N}_k]. \tag{8.33}$$

### *8.1.3 Use of a Penalty Function*

Aside from some special cases, such as those treated above, calculation of $\mathbf{Q}_k$ is complex, because evaluation of (8.18) to (8.23) requires a complicated program. All these calculations can be avoided by introducing a penalty function of the type

$$p(\boldsymbol{x}_{\mathrm{f}}) = \mu[\boldsymbol{\Gamma}\boldsymbol{x}_{\mathrm{f}} - \boldsymbol{a}]^{\mathrm{T}}\mathbf{A}[\boldsymbol{\Gamma}\boldsymbol{x}_{\mathrm{f}} - \boldsymbol{a}], \tag{8.34}$$

(where $\mathbf{A}$ is a positive definite matrix), since the optimal recursive system then begins with

$$\mathbf{Q}_0 = \mu\boldsymbol{\Gamma}^{\mathrm{T}}\mathbf{A}\boldsymbol{\Gamma}, \qquad \boldsymbol{\beta} = -\mathbf{Q}_0\,\boldsymbol{a}, \qquad \alpha = \boldsymbol{a}^{\mathrm{T}}\mathbf{Q}_0\,\boldsymbol{a}. \tag{8.35}$$

This approach can only be used, however, if a weight $\mu$ can be found which is large enough that the terminal condition is satisfied with sufficient precision, for all initial conditions of interest, and yet small enough that the optimal recurrence relations can be calculated with sufficient precision, taking account of the floating-point roundoff error present in the calculations. These two conditions are not always compatible, and determination of an appropriate $\mu$ is often delicate.

## 8.2 Minimum-Time Regulation with Bounded Energy

### *8.2.1 Definition of the Problem*

Let us consider a system described by (8.1), for which the total return $\mathscr{R}$, given by (8.2), corresponds to the energy consumed by the system in its evolution on the interval $[0, N]$. For simplicity, we shall now consider only the cases in which the final state is free, or the origin.

For an initial state $x_0$ and a given $N$, we have seen that there is a control law which minimizes $\mathscr{R}$, and that the minimum $\hat{\mathscr{R}}$ is quadratic in $x_0$. In the stationary case, if the matrices $Q$ are counted from the final state, they can be calculated sequentially using (8.6), and the minimum energy for which the terminal conditions can be satisfied will be

$$\hat{\mathscr{R}}(x_0, N) = x_0^{\mathrm{T}}\mathbf{Q}_N x_0 .$$

In case the available energy is limited to some maximum value $\gamma$, and the mission of the system is to be carried out in minimum time, a problem equivalent to the above results. The optimal time $N$ will be the smallest $n$ for which

$$\hat{\mathscr{R}}(x_0, m) \leq \gamma .$$

It is easy to see that $\hat{\mathscr{R}}(x_0, m)$ is a nonincreasing function of $m$. By definition,

$$\hat{\mathscr{R}}(x_0, m+1) \leq \mathscr{R}(x_0, m+1),$$

where $\mathscr{R}(x_0, m+1)$ is the return corresponding to some nonoptimal policy. Let us choose the particular nonoptimal policy, having $m + 1$ steps, that consists of the optimal policy for $m$ steps, followed by a zero control signal. The first $m$ controls result in a return $\hat{\mathscr{R}}(x_0, m)$ in steering the system to the origin. The following zero control results in an elementary return of zero, since $x = 0$ and $u = 0$, while the system remains at the origin. Thus this control policy also achieves the desired result (final state $= 0$), and

$$\mathscr{R}(x_0, m+1) = \hat{\mathscr{R}}(x_0, m).$$

Thus

$$\hat{\mathscr{R}}(x_0, m+1) \leq \hat{\mathscr{R}}(x_0, m). \tag{8.36}$$

### 8.2.2 *Practical Implementation*

Preliminary, off-line, solution of (8.5) and (8.6) supplies the matrices $\mathbf{L}_i$ and $\mathbf{Q}_i$, counted backward from the final time. At each sampling instant, use of these $\mathbf{Q}_i$, and the state at that instant, allows determination of an $N$ such that

$$x^{\mathrm{T}}\mathbf{Q}_N x \leq \gamma < x^{\mathrm{T}}\mathbf{Q}_{N+1} x . \tag{8.37}$$

The control

$$u = -\mathbf{L}_N x$$

is then applied to the system. Determination of $N$ is made easier by the non-increasing character of $\hat{\mathscr{R}}(\mathbf{x}, m)$.

This type of controller is illustrated by the flow chart of Fig. 8.3. This scheme adapts automatically to changes in the available energy. If there are no external changes made affecting the amount of energy available, the search for $N$ proceeds rapidly, in one step in principle.

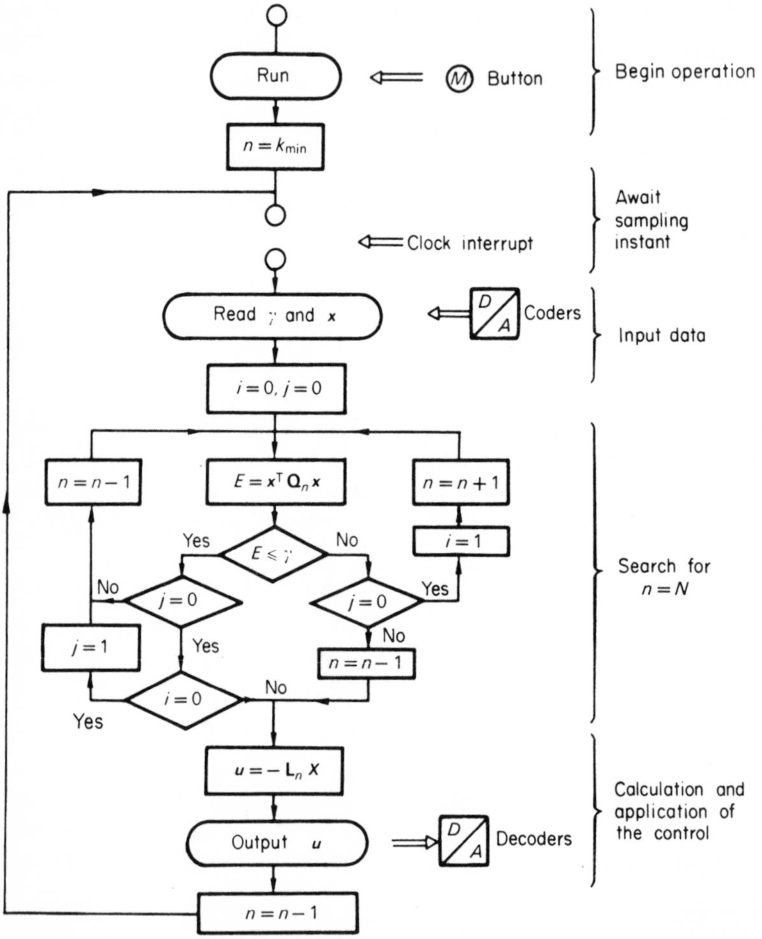


FIG. 8.3. Minimum-time control with bounded energy, where $j = 0 \rightarrow$ search with increasing $n$, $j = 1 \rightarrow$ search with decreasing $n$, and $i = 0 \rightarrow$ begin search.

## 8.3 Problems with Quadratic Constraints

We shall now consider quadratic constraints of the form

$$\mathscr{C}^i = \sum_{n=0}^{N-1} [x_n{}^{\mathrm{T}}\mathbf{P}_1{}^i x_n + 2x_n{}^{\mathrm{T}}\mathbf{P}_2{}^i u_n + u_n{}^{\mathrm{T}}\mathbf{P}_3{}^i u_n] \leq \gamma^i. \tag{8.38}$$

### 8.3.1 *Principle of Solution*

To simplify the presentation, we shall assume that only one constraint of the type (8.38) is present. The solution of this problem then proceeds by introducing a Lagrange multiplier $\lambda$ and considering the modified criterion

$$\begin{aligned}\mathscr{R}' = \mathscr{R} + \lambda\mathscr{C} &= \sum x^{\mathrm{T}}[\mathbf{R}_1 + \lambda\mathbf{P}_1]x + 2x^{\mathrm{T}}[\mathbf{R}_2 + \lambda\mathbf{P}_2]u + u^{\mathrm{T}}[\mathbf{R}_3 + \lambda\mathbf{P}_3]u \\ &= \sum x^{\mathrm{T}}\mathfrak{R}_1(\lambda)x + 2\,x^{\mathrm{T}}\mathfrak{R}_2(\lambda)u + u^{\mathrm{T}}\mathfrak{R}_3(\lambda)u,\end{aligned} \tag{8.39}$$

where we have defined

$$\mathfrak{R}_1(\lambda) = \mathbf{R}_1 + \lambda\mathbf{P}_1,$$

$$\mathfrak{R}_2(\lambda) = \mathbf{R}_2 + \lambda\mathbf{P}_2,$$

$$\mathfrak{R}_3(\lambda) = \mathbf{R}_3 + \lambda\mathbf{P}_3.$$

The solution is then carried out in three steps:

1. For various $\lambda$, solve the problem using the criterion $\mathscr{R}'$.
2. Determine whether the solution found with $\lambda = 0$ satisfies the constraint (8.38).
3. If the constraint is not satisfied for $\lambda = 0$, determine the value of $\lambda$ for which $\mathscr{C} = \gamma$, so that the constraint is satisfied with equality. The $\lambda$ thus determined should be positive in the case of search for a minimum.

The procedure needed in point 1 is identical to that in the unconstrained case, but must be carried out for each value of $\lambda$. Carrying out points 2 and 3 requires calculation of $\mathscr{C}$ for a given control policy. In the following section we shall show that $\mathscr{C}$ can be calculated recursively.

### 8.3.2 *Calculation of the Constraint*

Since the constraint is quadratic and the process and the control law are linear, $\mathscr{C}(x)$ will be a quadratic form:

$$\mathscr{C}(x) = x^{\mathrm{T}}\mathbf{W}x. \tag{8.40}$$

Using the control law

$$u = -\mathbf{L}x,$$

in (8.38), and counting time backward, yields

$$\mathscr{C}(x) = x^{\mathrm{T}}\mathbf{P}_1 x + 2x^{\mathrm{T}}\mathbf{P}_2(-\mathbf{L}x) + (-\mathbf{L}x)^{\mathrm{T}}\mathbf{P}_3(-\mathbf{L}x) + \mathscr{C}^-(x^-), \tag{8.41}$$

so that $\mathbf{W}$ in (8.40) is in fact

$$\mathbf{W} = \mathbf{P}_1 - (\mathbf{P}_2\mathbf{L})^{\mathrm{T}} - \mathbf{P}_2\mathbf{L} + \mathbf{L}^{\mathrm{T}}\mathbf{P}_3\mathbf{L} + (\mathbf{F} + \mathbf{HL})^{\mathrm{T}}\mathbf{W}^-(\mathbf{F} + \mathbf{HL}). \tag{8.42}$$

Relation (8.42) is a matrix recurrence relation, to be initialized with $\mathbf{W} = \mathbf{0}$. Thus the matrices $\mathbf{W}$ can be calculated recursively at the same time as $\mathbf{L}$ and $\mathbf{Q}$, using (8.42).

### *8.3.3 Determination of the Lagrange Multiplier*

In solving (8.5), (8.6), and (8.42) for various values of $\lambda$, the matrices $\mathbf{W}_N(\lambda)$ are obtained for $N$. For some initial value $X$, it then suffices to determine $\lambda$ such that

$$X^{\mathrm{T}}\mathbf{W}_N(\lambda)X = \gamma. \tag{8.43}$$

This value of $\lambda$ can be found by interpolation.

### *8.3.4 Solution in Practice*

In practice, the solution is carried out in two parts:

1. An off-line preliminary calculation of the $\mathbf{W}_n(\lambda_i)$ and $\mathbf{L}_n(\lambda_i)$, using the relations (8.5), (8.6), and (8.42), and diagrammed in Fig. 8.4.
2. A real-time computation during system operation, involving determination of the proper $\lambda$ to be used, and formation of the control signal at each cycle of system operation. The flow chart of Fig. 8.5 indicates the controller operation.

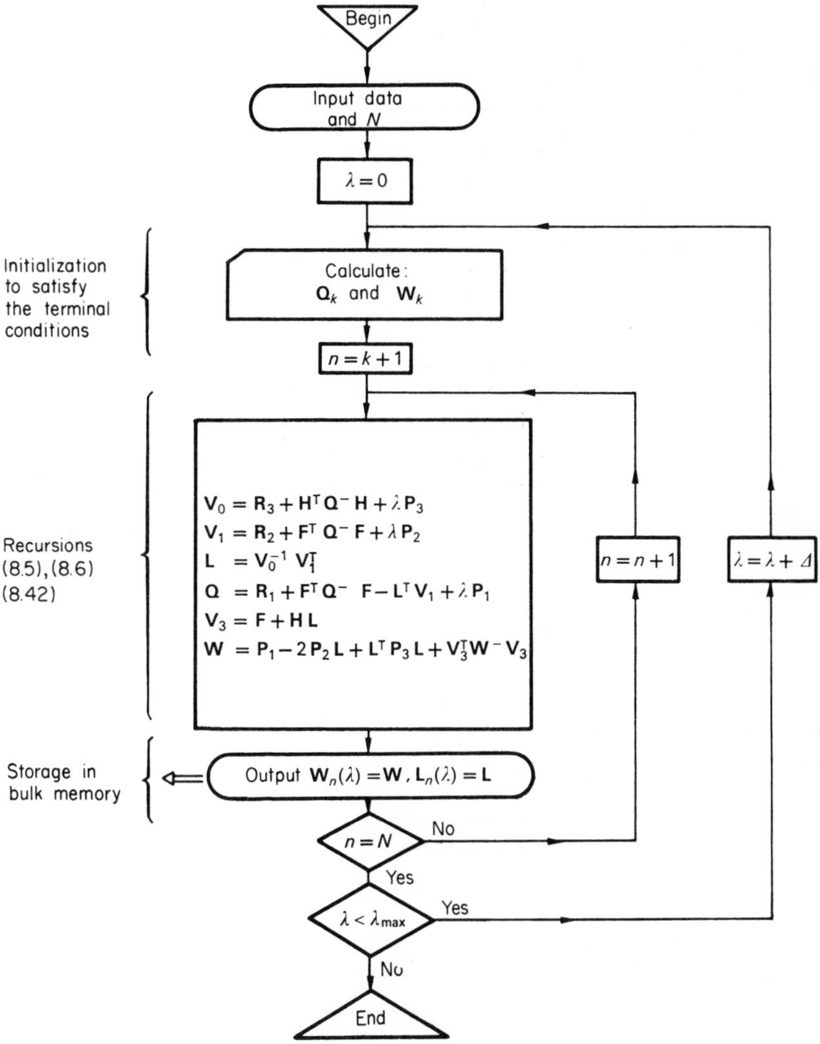


FIG. 8.4. Controller synthesis.

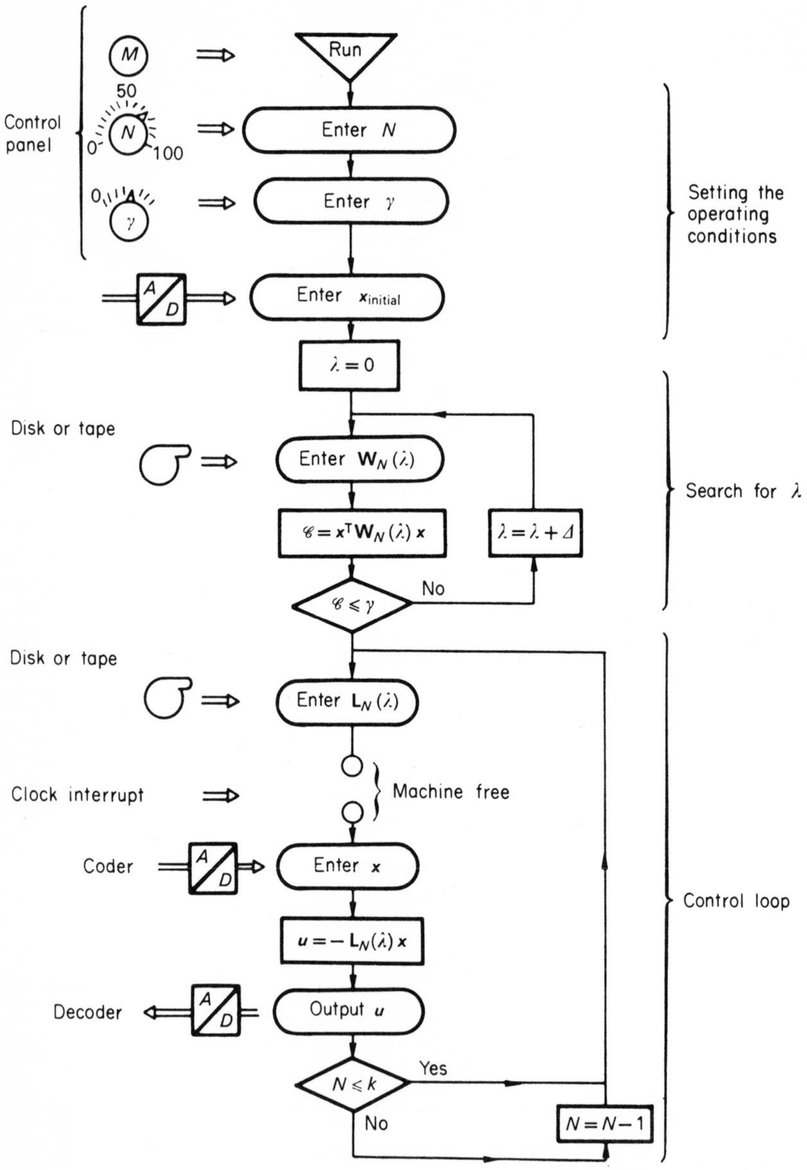


FIG. 8.5. Optimal controller with quadratic constraints.

## 8.4 The Case of an Infinite Horizon

We shall assume here that the process and criterion are stationary.

### *8.4.1 Theoretical Aspects*

In considering the sense of the problem when $N \to \infty$, three cases can be distinguished, intuitively:

a. As $N \to \infty$, $\mathbf{Q} \to \mathbf{Q}_\infty = \mathbf{0}$ and $\mathbf{L} \to \mathbf{L}_\infty = \mathbf{0}$ (e.g., the case of a criterion depending only on $\boldsymbol{u}$). The optimal control action is then to allow the system to evolve uncontrolled toward the origin (if the system is stable), or toward infinity (if the system is unstable).

b. As $N \to \infty$, $\mathbf{Q}_\infty$ and $\mathbf{L}_\infty$ converge to certain finite, nonzero matrices. This is the interesting case, and results when the criterion involves both $\boldsymbol{x}$ and $\boldsymbol{u}$, and the elementary return vanishes only when $\boldsymbol{x}$ satisfies the terminal conditions.

c. As $N \to \infty$, $\mathbf{Q} \to \infty$. In these problems the system objective can not be reached, even in infinite time, with a finite cost.

Usually, when the horizon tends to infinity, that the criterion remains finite automatically ensures fulfilment of the terminal condition. The terminal condition can then simply be omitted, and the free end point problem is recovered.

If the problem retains its sense as $N \to \infty$, we have seen in Chapter 3 that the criterion and the control law become stationary. The matrix appearing in the quadratic form for the optimal return then satisfies

$$\mathbf{Q}_\infty = \mathbf{R}_1 + \mathbf{F}^T\mathbf{Q}_\infty\mathbf{F} - [\mathbf{R}_2 + \mathbf{F}^T\mathbf{Q}_\infty\mathbf{H}][\mathbf{R}_3 + \mathbf{H}^T\mathbf{Q}_\infty\mathbf{H}]^{-1} \times [\mathbf{R}_2{}^T + \mathbf{H}^T\mathbf{Q}_\infty\mathbf{F}], \tag{8.44}$$

as follows from (8.6). This is an implicit nonlinear matrix equation. It is best solved iteratively, which is to say that $\mathbf{Q}_\infty$ is taken as the limit of the matrices $\mathbf{Q}$ calculated using (8.6). Since the problem with terminal constraints and the free end point problem both correspond to the same limit problem, the solution of (8.6) can be started with $\mathbf{Q} = \mathbf{0}$, which simplifies the calculations. In fact, if the sequence calculated by (8.6) is uniformly convergent, an arbitrary $\mathbf{Q}$ can be used for initialization.

### *8.4.2 Practical Aspects*

Even though an infinite horizon may pose theoretical problems relating to convergence, from the practical point of view its treatment is simple, since the control law is stationary.

In case constraints are present, the matrices $\mathbf{W}_N(\lambda)$ are also stationary, and it is only necessary to store matrices $\mathbf{W}(\lambda)$ for various values of $\lambda$. The flow chart of the controller is then the same as shown in Fig. 8.5. However, it may in this case be feasible to store the $\mathbf{W}(\lambda)$ in the fast working memory. This allows easy recalculation of $\lambda$ at each sampling instant, and thus ensures optimal regulation even with an energy bound which may change in the course of system operation.

**Remark.** The infinite-horizon problem can also be solved by the method of Wiener, extended to the sampled, multivariable case (see Boudarel *et al.* [1, Vol. 4]). However, practical implementation of that solution involves serious numerical difficulties. With the approach used here, these difficulties do not arise, since the algorithms used are very well adapted to digital calculation.

# Chapter 9 Numerical Optimal Control of a Stochastic Process

We shall begin by considering a control problem analogous to that of the preceding chapter, but in which a random noise term enters into the equation describing the evolution of the process. The state will again be assumed to be completely measurable, so that the output vector will be just the state vector, with no measurement noise. We shall show that in this case the control law is the same as if no process disturbances were present.

In the second section we return to the case of a system without process perturbations, but now assume that it is possible for a control signal to be missed. That is, we assume that for one reason or another, no control action is applied to the system at some sampling time. In this case the control law is not changed, but the matrix recursion relations become more complicated.

The third section takes up the most general problem, in which process perturbations are again present, but in which the state vector is not completely measurable. We shall establish the important separation theorem, which states that it is sufficient in this case to adjoin to the controller, constructed for the noiseless case, an element which estimates the state vector based on whatever information is available.

In this chapter, we shall use the convention that when the time index is omitted, time is to be counted backward from the final instant.

## 9.1 Completely Measurable Processes

### *9.1.1 The Optimal Recurrence Equation*

The system to be considered is

$$\boldsymbol{x}_{n+1} = \mathbf{F}_n \boldsymbol{x}_n + \mathbf{H}_n \boldsymbol{u}_n + \mathbf{\Pi}_n \boldsymbol{\Delta}_n , \tag{9.1}$$

$$\boldsymbol{i}_n = \boldsymbol{x}_n . \tag{9.2}$$

Thus, the information vector $\boldsymbol{i}_n$ is just the state vector itself. The random vector $\boldsymbol{\Delta}_n$ perturbing the process is such that

$$\mathrm{E}[\boldsymbol{\Delta}_n] = \boldsymbol{0} \tag{9.3}$$

$$\mathrm{E}[\boldsymbol{\Delta}_n \boldsymbol{\Delta}_m{}^{\mathrm{T}}] = \begin{cases} \mathbf{0}, & n \neq m \\ \mathbf{\Phi}, & n = m . \end{cases} \tag{9.4}$$

The quantity to be minimized is now

$$\mathrm{E}[\mathscr{R}] = C(\boldsymbol{x}) .$$

The attained minimum $\hat{C}(\boldsymbol{x})$ must satisfy the optimality equation

$$\hat{C}(\boldsymbol{x}) = \min_{\boldsymbol{u}} \mathrm{E}[\boldsymbol{x}^{\mathrm{T}}\mathbf{R}_1\boldsymbol{x} + 2\boldsymbol{x}^{\mathrm{T}}\mathbf{R}_2 \boldsymbol{u} + \boldsymbol{u}^{\mathrm{T}}\mathbf{R}_3 \boldsymbol{u} + \hat{C}^{-}(\boldsymbol{x}^{-})] . \tag{9.5}$$

It is clear that, if the horizon is zero,

$$\hat{C}(\boldsymbol{x}) = 0 .$$

Since this equation is the same as in the deterministic case, it can be assumed that $\hat{C}(\boldsymbol{x})$ depends quadratically on $\boldsymbol{x}$. The presence of the noise $\boldsymbol{\Delta}$, however, introduces additional terms corresponding to the return relative to the case of a final state $\boldsymbol{x} \neq \boldsymbol{0}$, since the final state can not be zero because of the noise.

### *9.1.2 Solution of the Recurrence Equation*

Letting

$$\hat{C}(\boldsymbol{x}) = \boldsymbol{x}^{\mathrm{T}}\mathbf{Q}\boldsymbol{x} + k ,$$

Eq. (9.5) becomes

$$k + \boldsymbol{x}^{\mathrm{T}}\mathbf{Q}\boldsymbol{x} = \min_{\boldsymbol{u}}[\boldsymbol{x}^{\mathrm{T}}\mathbf{R}_1\boldsymbol{x} + 2\boldsymbol{x}^{\mathrm{T}}\mathbf{R}_2\boldsymbol{u} + \boldsymbol{u}^{\mathrm{T}}\mathbf{R}_3\boldsymbol{u} + \mathrm{E}\{[\mathbf{F}\boldsymbol{x} + \mathbf{H}\boldsymbol{u} + \boldsymbol{\Pi}\Delta]^{\mathrm{T}}\mathbf{Q}^{-}[\mathbf{F}\boldsymbol{x} + \mathbf{H}\boldsymbol{u} + \boldsymbol{\Pi}\Delta] + k^{-}\}] \tag{9.6}$$

$$= \min_{\boldsymbol{u}}\{\boldsymbol{x}^{\mathrm{T}}\mathbf{R}_1\boldsymbol{x} + 2\boldsymbol{x}^{\mathrm{T}}\mathbf{R}_2\boldsymbol{u} + \boldsymbol{u}^{\mathrm{T}}\mathbf{R}_3\boldsymbol{u} + [\mathbf{F}\boldsymbol{x} + \mathbf{H}\boldsymbol{u}]^{\mathrm{T}}\mathbf{Q}^{-}[\mathbf{F}\boldsymbol{x}+\mathbf{H}\boldsymbol{u}]\} + \mathrm{E}\{\Delta^{\mathrm{T}}\boldsymbol{\Pi}^{\mathrm{T}}\mathbf{Q}^{-}\boldsymbol{\Pi}\Delta\} + k^{-}.$$

Here the term under the minimization operator is identical to that in the corresponding deterministic case, and leads to the linear control law

$$\hat{\boldsymbol{u}} = -\mathbf{L}\boldsymbol{x}. \tag{9.7}$$

Using the fact that

$$\mathrm{trace}(\mathbf{AB}) = \mathrm{trace}(\mathbf{BA}) \Rightarrow \boldsymbol{x}^{\mathrm{T}}\boldsymbol{y} = \mathrm{trace}(\boldsymbol{y}\boldsymbol{x}^{\mathrm{T}}),$$

the term under the expectation operator can be written

$$\mathrm{trace}[\boldsymbol{\Pi}^{\mathrm{T}}\mathbf{Q}^{-}\boldsymbol{\Pi}\Delta\Delta^{\mathrm{T}}] \rightarrow \mathrm{trace}[\boldsymbol{\Pi}^{\mathrm{T}}\mathbf{Q}^{-}\boldsymbol{\Pi}\boldsymbol{\Phi}].$$

Thus, solution of (9.6) leads to the system of recursions

$$\mathbf{L} = [\mathbf{R}_3 + \mathbf{H}^{\mathrm{T}}\mathbf{Q}^{-}\mathbf{H}]^{-1}[\mathbf{R}_2{}^{\mathrm{T}} + \mathbf{H}^{\mathrm{T}}\mathbf{Q}^{-}\mathbf{F}], \tag{9.8}$$

$$\mathbf{Q} = \mathbf{R}_1 + \mathbf{F}^{\mathrm{T}}\mathbf{Q}^{-}\mathbf{F} - [\mathbf{R}_2 + \mathbf{F}^{\mathrm{T}}\mathbf{Q}^{-}\mathbf{H}]\mathbf{L}, \tag{9.9}$$

$$k = k^{-} + \mathrm{trace}[\boldsymbol{\Pi}^{\mathrm{T}}\mathbf{Q}^{-}\boldsymbol{\Pi}\boldsymbol{\Phi}]. \tag{9.10}$$

### *9.1.3 Interpretation of the Results*

Relations (9.8) and (9.9) are independent of the third relation, (9.10), and are identical to the relations obtained in the deterministic case. Thus the optimal control law (9.7) is unchanged by the perturbation $\Delta$, and the only effect is to degrade the value of the cost function by the term $k$. This result is easily understood by considering that the $\Delta_n$ are independent zero-mean vectors, and hence it is natural to proceed as if $\Delta_n$ were zero and the problem deterministic.

## 9.2 The Case of Possible Missed Controls

### 9.2.1 *Framework of the Problem*

The controller has the task of constructing a control $\boldsymbol{u}_n$ based on information about the process state $\boldsymbol{x}_n$. For various reasons, this control may not be forthcoming. Calculation of the control may be interrupted for execution of some other program of higher priority in the computing system; it may not be possible to execute the computed control; etc. We shall then say that the controller is subject to missed controls. The missed control can be replaced either with a zero control signal, or with the last available control, if there exists a register for storing the control over the following sampling period.

We shall see that if missed controls are assumed to be independent random events, occurring with probability $q = 1 - p$, where $p$ is the probability of correct operation, then the optimal control law can be found. The process and the criterion are assumed to have a probability $p$ of evolving normally, and a probability $q$ of evolving according to some strategy called into play if a control signal is missed. This type of system was examined in Section 5.2.4, where it was found that the control law was again linear, and the optimal return quadratic. We shall derive the formulas for the two specific strategies mentioned above.

### 9.2.2 *Use of the Zero Control Signal*

In use of the zero control signal the two possible laws under which the process and criterion evolve are

$$\text{with probability } p: \quad \begin{aligned} \boldsymbol{x}^- &= \mathbf{F}\boldsymbol{x} + \mathbf{H}\boldsymbol{u}, \\ r &= \boldsymbol{x}^{\mathrm{T}}\mathbf{R}_1\boldsymbol{x} + 2\boldsymbol{x}^{\mathrm{T}}\mathbf{R}_2\boldsymbol{u} + \boldsymbol{u}^{\mathrm{T}}\mathbf{R}_3\boldsymbol{u}; \end{aligned}$$

$$\text{with probability } q = 1 - p: \quad \begin{aligned} \boldsymbol{x}^- &= \mathbf{F}\boldsymbol{x}, \\ r &= \boldsymbol{x}^{\mathrm{T}}\mathbf{R}_1\boldsymbol{x}. \end{aligned}$$

Using the formulas of Section 5.2.4, these result in

$$\begin{aligned} \mathbf{L} &= [p(\mathbf{R}_3 + \mathbf{H}^{\mathrm{T}}\mathbf{Q}^-\mathbf{H})]^{-1}[p(\mathbf{R}_2{}^{\mathrm{T}} + \mathbf{H}^{\mathrm{T}}\mathbf{Q}^-\mathbf{F})] \\ &= (\mathbf{R}_3 + \mathbf{H}^{\mathrm{T}}\mathbf{Q}^-\mathbf{H})^{-1}(\mathbf{R}_2{}^{\mathrm{T}} + \mathbf{H}^{\mathrm{T}}\mathbf{Q}^-\mathbf{F}), \end{aligned} \tag{9.11}$$

$$\mathbf{Q} = \mathbf{R}_1 + \mathbf{F}^{\mathrm{T}}\mathbf{Q}^-\mathbf{F} - p(\mathbf{R}_2 + \mathbf{F}^{\mathrm{T}}\mathbf{Q}^-\mathbf{H})(\mathbf{R}_3 + \mathbf{H}^{\mathrm{T}}\mathbf{Q}^-\mathbf{H})^{-1}(\mathbf{R}_2{}^{\mathrm{T}} + \mathbf{H}^{\mathrm{T}}\mathbf{Q}^-\mathbf{F}). \tag{9.12}$$

Only the equation for $\mathbf{Q}$ is affected by $p$, and only in the last term.

### 9.2.3 *Use of the Last Control Signal*

Use of the previous control signal corresponds to the introduction of delay into the system, and the vector

$$\boldsymbol{y} = \begin{bmatrix} \boldsymbol{x} \\ \boldsymbol{v} \end{bmatrix}$$

should be used as the total state vector, where $\boldsymbol{v}$ is the previous control. The two possible modes of evolution are

with probability $p$:
$$\begin{bmatrix} \boldsymbol{x} \\ \boldsymbol{v} \end{bmatrix}^{-} = \begin{bmatrix} \mathbf{F} & \mathbf{0} \\ \mathbf{0} & \mathbf{0} \end{bmatrix} \begin{bmatrix} \boldsymbol{x} \\ \boldsymbol{v} \end{bmatrix} + \begin{bmatrix} \mathbf{H} \\ \mathbf{1} \end{bmatrix} \boldsymbol{u}, \tag{9.13}$$

$$r = \begin{bmatrix} \boldsymbol{x} \\ \boldsymbol{v} \end{bmatrix}^{\mathrm{T}} \begin{bmatrix} \mathbf{R}_1 & \mathbf{0} \\ \mathbf{0} & \mathbf{0} \end{bmatrix} \begin{bmatrix} \boldsymbol{x} \\ \boldsymbol{v} \end{bmatrix} + 2 \begin{bmatrix} \boldsymbol{x} \\ \boldsymbol{v} \end{bmatrix}^{\mathrm{T}} \begin{bmatrix} \mathbf{Q}_2 \\ \mathbf{0} \end{bmatrix} \boldsymbol{u} + \boldsymbol{u}^{\mathrm{T}} \mathbf{R}_3 \boldsymbol{u}; \tag{9.14}$$

with probability $1 - p = q$:
$$\begin{bmatrix} \boldsymbol{x} \\ \boldsymbol{v} \end{bmatrix}^{-} = \begin{bmatrix} \mathbf{F} & \mathbf{H} \\ \mathbf{0} & \mathbf{1} \end{bmatrix} \begin{bmatrix} \boldsymbol{x} \\ \boldsymbol{v} \end{bmatrix}, \tag{9.15}$$

$$r = \begin{bmatrix} \boldsymbol{x} \\ \boldsymbol{v} \end{bmatrix}^{\mathrm{T}} \begin{bmatrix} \mathbf{R}_1 & \mathbf{R}_2 \\ \mathbf{R}_2^{\mathrm{T}} & \mathbf{R}_3 \end{bmatrix} \begin{bmatrix} \boldsymbol{x} \\ \boldsymbol{v} \end{bmatrix}. \tag{9.16}$$

Taking account of the partitioning of $\boldsymbol{y}$, $\mathbf{Q}$ should be written

$$\mathbf{Q} = \begin{bmatrix} \mathbf{Q}_1 & \mathbf{Q}_2 \\ \mathbf{Q}_2^{\mathrm{T}} & \mathbf{Q}_3 \end{bmatrix}. \tag{9.17}$$

Again using the formulas of Section 5.2.4 yields

$$\begin{aligned} \mathbf{L} &= \left\{ p \left( \mathbf{R}_3 + \begin{bmatrix} \mathbf{H} \\ \mathbf{1} \end{bmatrix}^{\mathrm{T}} \begin{bmatrix} \mathbf{Q}_1 & \mathbf{Q}_2 \\ \mathbf{Q}_2^{\mathrm{T}} & \mathbf{Q}_3 \end{bmatrix}^{-} \begin{bmatrix} \mathbf{H} \\ \mathbf{1} \end{bmatrix} \right) \right\}^{-1} \\ &\quad \times \left\{ p \left( \begin{bmatrix} \mathbf{R}_2 \\ \mathbf{0} \end{bmatrix}^{\mathrm{T}} + \begin{bmatrix} \mathbf{H} \\ \mathbf{1} \end{bmatrix}^{\mathrm{T}} \begin{bmatrix} \mathbf{Q}_1 & \mathbf{Q}_2 \\ \mathbf{Q}_2^{\mathrm{T}} & \mathbf{Q}_3 \end{bmatrix}^{-} \begin{bmatrix} \mathbf{F} & \mathbf{0} \\ \mathbf{0} & \mathbf{0} \end{bmatrix} \right) \right\} \\ &= [\mathbf{R}_3 + \mathbf{H}^{\mathrm{T}} \mathbf{Q}_1^{-} \mathbf{H} + \mathbf{Q}_2^{-\mathrm{T}} \mathbf{H} + \mathbf{H}^{\mathrm{T}} \mathbf{Q}_2^{-} + \mathbf{Q}_3^{-}]^{-1} \\ &\quad \times [\mathbf{R}_2^{\mathrm{T}} + (\mathbf{H}^{\mathrm{T}} \mathbf{Q}_1^{-} + \mathbf{Q}_2^{-}) \mathbf{F}, \mathbf{0}]. \end{aligned} \tag{9.18}$$

Note that, as a result, $\boldsymbol{u}$ depends only on $\boldsymbol{x}$, and not on $\boldsymbol{v}$.

Defining **A** and **B** in the fashion obvious from (9.18) then

$$\mathbf{L} = \mathbf{A}[\mathbf{B}, \mathbf{0}], \tag{9.19}$$

and the expression for **Q** becomes

$$\begin{bmatrix} \mathbf{Q}_1 & \mathbf{Q}_2 \\ \mathbf{Q}_2^{\mathrm{T}} & \mathbf{Q}_3 \end{bmatrix} = \begin{bmatrix} \mathbf{R}_1 & q\mathbf{R}_2 \\ q\mathbf{R}_2^{\mathrm{T}} & q\mathbf{R}_3 \end{bmatrix} + \left\{ p \begin{bmatrix} \mathbf{F}^{\mathrm{T}} & \mathbf{0} \\ \mathbf{0} & \mathbf{0} \end{bmatrix} \begin{bmatrix} \mathbf{Q}_1 & \mathbf{Q}_2 \\ \mathbf{Q}_2^{\mathrm{T}} & \mathbf{Q}_3 \end{bmatrix}^{-} \begin{bmatrix} \mathbf{F} & \mathbf{0} \\ \mathbf{0} & \mathbf{0} \end{bmatrix} \right.$$
$$\left. + q \begin{bmatrix} \mathbf{F}^{\mathrm{T}} & \mathbf{0} \\ \mathbf{H}^{\mathrm{T}} & \mathbf{1} \end{bmatrix} \begin{bmatrix} \mathbf{Q}_1 & \mathbf{Q}_2 \\ \mathbf{Q}_2^{\mathrm{T}} & \mathbf{Q}_3 \end{bmatrix}^{-} \begin{bmatrix} \mathbf{F} & \mathbf{H} \\ \mathbf{0} & \mathbf{1} \end{bmatrix} \right\}$$
$$- p[\mathbf{B}, \mathbf{0}]^{\mathrm{T}} \mathbf{A} [\mathbf{B}, \mathbf{0}].$$

The recurrences for $\mathbf{Q}_1$, $\mathbf{Q}_2$, and $\mathbf{Q}_3$ are

$$\mathbf{Q}_1 = \mathbf{R}_1 + \mathbf{F}^{\mathrm{T}}\mathbf{Q}_1^{-}\mathbf{F} - p[\mathbf{R}_2 + \mathbf{F}^{\mathrm{T}}(\mathbf{Q}_1^{-}\mathbf{H} + \mathbf{Q}_2^{-\mathrm{T}})][\mathbf{R}_3 + \mathbf{H}^{\mathrm{T}}\mathbf{Q}_1^{-}\mathbf{H}$$
$$+ \mathbf{H}^{\mathrm{T}}\mathbf{Q}_2^{-} + \mathbf{Q}_2^{-\mathrm{T}}\mathbf{H} + \mathbf{Q}_3^{-}]^{-1}[\mathbf{R}_2^{\mathrm{T}} + (\mathbf{H}^{\mathrm{T}}\mathbf{Q}_1^{-} + \mathbf{Q}_2^{-})\mathbf{F}], \tag{9.20}$$

$$\mathbf{Q}_2 = q(\mathbf{R}_2 + \mathbf{F}^{\mathrm{T}}\mathbf{Q}_1^{-}\mathbf{H} + \mathbf{F}^{\mathrm{T}}\mathbf{Q}_2^{-}), \tag{9.21}$$

$$\mathbf{Q}_3 = q(\mathbf{R}_3 + \mathbf{H}^{\mathrm{T}}\mathbf{Q}_1^{-}\mathbf{H} + \mathbf{H}^{\mathrm{T}}\mathbf{Q}_2^{-} + \mathbf{Q}_2^{-\mathrm{T}}\mathbf{H} + \mathbf{Q}_3^{-}). \tag{9.22}$$

Since there are no terminal conditions, $\mathbf{Q}_1$, $\mathbf{Q}_2$, and $\mathbf{Q}_3$ are initialized with zero.

It might be asked how $\boldsymbol{v}$ should be initialized, if that is possible, before actual operation of the system (before state measurements begin). Let us consider the case in which the initial state $X$ is random, and described by some *a priori* distribution. The mathematical expectation of the return is then

$$\begin{aligned} \mathscr{C} &= \mathrm{E}[X^{\mathrm{T}}\mathbf{Q}_1 X + 2X^{\mathrm{T}}\mathbf{Q}_2\boldsymbol{v} + \boldsymbol{v}^{\mathrm{T}}\mathbf{Q}_3\boldsymbol{v}] \\ &= \mathrm{trace}[\mathbf{Q}_1\boldsymbol{\Phi}_X] + 2\bar{X}^{\mathrm{T}}\mathbf{Q}_2\boldsymbol{v} + \boldsymbol{v}^{\mathrm{T}}\mathbf{Q}_3\boldsymbol{v}, \end{aligned} \tag{9.23}$$

where

$$\boldsymbol{\Phi}_X = \mathrm{E}[XX^{\mathrm{T}}], \qquad \bar{X} = \mathrm{E}[X]. \tag{9.24}$$

The $\boldsymbol{v}$ which minimizes $\mathscr{C}$ is

$$\hat{\boldsymbol{v}} = -\mathbf{Q}_3^{-1}\mathbf{Q}_2\bar{X}, \tag{9.25}$$

to which there corresponds

$$\hat{\mathscr{C}} = \mathrm{trace}[\mathbf{Q}_1\boldsymbol{\Phi}_X] - \bar{X}^{\mathrm{T}}\mathbf{Q}_2\mathbf{Q}_3^{-1}\mathbf{Q}_2\bar{X}. \tag{9.26}$$

The first term of this is the average return if $\boldsymbol{v}$ is initialized at zero, and the second term represents the improvement obtained with optimal initialization.

## 9.3 Processes with a State Not Completely Measurable

In general, the state is not completely measurable, but is known only through some data set $\boldsymbol{i}_0, \boldsymbol{i}_1, \ldots, \boldsymbol{i}_n$. In this case, as we have seen in Chapter 5, the optimal control functions and the optimal return are functions of $\boldsymbol{i}_0, \ldots, \boldsymbol{i}_n$. This dependence is a serious obstacle, since the number of variables entering into the control and return functions then increases with increasing $n$. However, as indicated earlier, it is sometimes possible to replace $\boldsymbol{i}_0, \ldots, \boldsymbol{i}_n$ by a vector $\hat{\boldsymbol{x}}_n$, an estimate of the state $\boldsymbol{x}$, conditioned by the data $\boldsymbol{i}_0, \ldots, \boldsymbol{i}_n$. We shall establish this result in the present case, and we shall show that the optimal control law is the same as that in the deterministic case, with measurable state, except that the estimate $\hat{\boldsymbol{x}}$ is used in place of $\boldsymbol{x}$. This important result is known as the separation theorem.

### *9.3.1 Preliminary Formulas*

To represent mathematical expectation conditioned by the data available at time $n$, we shall use the notation

$$\mathrm{E}|_n[\cdots], \qquad \text{or} \quad \mathrm{E}|[\;].$$

With the index convention and backward time progression used, the notation $\mathrm{E}|_+[\cdots]$ then means mathematical expectation conditioned by data available at the previous (real) time, $n-1$.

Let $\boldsymbol{x}$ be the process state vector, and let

$$\boldsymbol{x} = \hat{\boldsymbol{x}} + \boldsymbol{\varepsilon}, \tag{9.27}$$

where

$$\hat{\boldsymbol{x}} = \mathrm{E}|[\boldsymbol{x}] \Rightarrow \mathrm{E}|\,[\boldsymbol{\varepsilon}] = \boldsymbol{0}. \tag{9.28}$$

Also let

$$\mathrm{E}|[\boldsymbol{\varepsilon}\boldsymbol{\varepsilon}^{\mathrm{T}}] = \boldsymbol{\Psi}. \tag{9.29}$$

It is then easy to see that

$$\begin{aligned} \mathrm{E}|[\boldsymbol{x}^{\mathrm{T}}\mathbf{Q}\boldsymbol{x}] &= \mathrm{E}|[(\hat{\boldsymbol{x}} + \boldsymbol{\varepsilon})^{\mathrm{T}}\mathbf{Q}(\hat{\boldsymbol{x}} + \boldsymbol{\varepsilon})] \\ &\hat{\boldsymbol{x}}^{\mathrm{T}}\mathbf{Q}\hat{\boldsymbol{x}} + \mathrm{trace}[\mathbf{Q}\boldsymbol{\Psi}]. \end{aligned} \tag{9.30}$$

Use of the process equation

$$x^{-} = \mathbf{F}x + \mathbf{H}u + \mathbf{\Pi}\Delta$$

leads easily to

$$\begin{aligned}
\mathrm{E}|[x^{-}] &= \mathbf{F}\hat{x} + \mathbf{H}u\,, \\
\mathrm{E}|[x^{-\mathrm{T}}\mathbf{Q}x^{-}] &= [\mathbf{F}\hat{x} + \mathbf{H}u]^{\mathrm{T}}\mathbf{Q}[\mathbf{F}\hat{x} + \mathbf{H}u] \\
&\quad + \mathrm{E}|\,[(\mathbf{F}\varepsilon)^{\mathrm{T}}\mathbf{Q}(\mathbf{F}\varepsilon) + \Delta^{\mathrm{T}}\mathbf{\Pi}^{\mathrm{T}}\mathbf{Q}\mathbf{\Pi}\Delta] \\
&= [\mathbf{F}\hat{x} + \mathbf{H}u]^{\mathrm{T}}\mathbf{Q}[\mathbf{F}\hat{x} + \mathbf{H}u] \\
&\quad + \mathrm{trace}[\mathbf{F}^{\mathrm{T}}\mathbf{Q}\mathbf{F}\mathbf{\Psi}] + \mathrm{trace}[\mathbf{\Pi}^{\mathrm{T}}\mathbf{Q}\mathbf{\Pi}\mathbf{\Phi}]. \qquad (9.31)
\end{aligned}$$

### 9.3.2 *The Separation Theorem*

The optimal return is a function $\hat{\mathscr{R}}(i_0, \ldots, i_n)$, which, by application of the principle of optimality, satisfies

$$\hat{\mathscr{R}} = \operatorname*{opt}_{u} \mathrm{E}|[x^{\mathrm{T}}\mathbf{R}_1 x + 2\,x^{\mathrm{T}}\mathbf{R}_2\,u + u^{\mathrm{T}}\mathbf{R}_3\,u + \hat{\mathscr{R}}^{-}].$$

Taking account of (9.30), this becomes

$$\hat{\mathscr{R}} = \operatorname*{opt}_{u}\{\hat{x}^{\mathrm{T}}\mathbf{R}_1\hat{x} + 2\hat{x}^{\mathrm{T}}\mathbf{R}_2\,u + u^{\mathrm{T}}\mathbf{R}_3\,u + \mathrm{trace}[\mathbf{R}_1\mathbf{\Psi}] + \mathrm{E}|[\hat{\mathscr{R}}^{-}]\}. \qquad (9.32)$$

Since

$$\hat{\mathscr{R}}_0 \equiv 0,$$

$\hat{\mathscr{R}}_1$ will be a quadratic form in $x_1$. As in the deterministic case, let us assume this result is true for all $m$, and then establish this hypothesis recursively, using

$$\hat{\mathscr{R}} = \hat{x}^{\mathrm{T}}\mathbf{Q}\hat{x} + k. \qquad (9.33)$$

Expanding the last term of (9.32) yields

$$\begin{aligned}
\mathrm{E}|[\hat{\mathscr{R}}^{-}] &= \mathrm{E}|[\hat{x}^{-\mathrm{T}}\mathbf{Q}^{-}\hat{x}^{-} + k^{-}] \\
&= \mathrm{E}|\{\mathrm{E}|_{-}[x^{-\mathrm{T}}\mathbf{Q}^{-}x^{-}] - \mathrm{trace}[\mathbf{Q}^{-}\mathbf{\Phi}^{-}]\} + k^{-} \\
&= \mathrm{E}|[x^{-\mathrm{T}}\mathbf{Q}^{-}x^{-}] - \mathrm{E}|\{\mathrm{trace}[\mathbf{Q}^{-}\mathbf{\Phi}^{-}]\} + k^{-} \qquad (9.34)
\end{aligned}$$

using (9.30) and the fact that

$$\mathrm{E}|\{\mathrm{E}|_{-}[\cdots]\} = \mathrm{E}|[\cdots].$$

Thus, using (9.31) and substituting the result into (9.32), the quadratic and constant terms can be separated:

$$\hat{\boldsymbol{x}}^{\mathrm{T}}\mathbf{Q}\hat{\boldsymbol{x}} = \mathrm{opt}[\hat{\boldsymbol{x}}^{\mathrm{T}}\mathbf{R}_1\hat{\boldsymbol{x}} + 2\hat{\boldsymbol{x}}^{\mathrm{T}}\mathbf{R}_2\boldsymbol{u} + \boldsymbol{u}^{\mathrm{T}}\mathbf{R}_3\boldsymbol{u} + (\mathbf{F}\hat{\boldsymbol{x}} + \mathbf{H}\boldsymbol{u})^{\mathrm{T}}\mathbf{Q}^{-}(\mathbf{F}\boldsymbol{x} + \mathbf{H}\boldsymbol{u})], \tag{9.35}$$

$$k = k^{-} + \mathrm{trace}[\mathbf{R}_1\boldsymbol{\Psi}] + \mathrm{trace}[\mathbf{F}^{\mathrm{T}}\mathbf{Q}^{-}\mathbf{F}\boldsymbol{\Psi}] + \mathrm{trace}[\boldsymbol{\Pi}^{\mathrm{T}}\mathbf{Q}^{-}\boldsymbol{\Pi}\boldsymbol{\Phi}] - \mathrm{E}|[\mathbf{Q}^{-}\boldsymbol{\Psi}^{-}]. \tag{9.36}$$

The right side of (9.35) is the same expression as obtained in the deterministic case. Thus the solutions are the same, and the matrices $\mathbf{Q}$ are identical to those obtained in the deterministic case, with free final state [formulas (9.8) and (9.9)]. This establishes the separation theorem.

**Separation Theorem.** For a stochastic process with a state which is not exactly measurable, the control law relating $\hat{\boldsymbol{u}}$ to $\hat{\boldsymbol{x}}$ is linear, and identical to that for the deterministic problem with measurable state. The control problem thus separates into two independent parts: (1) an optimal control problem for a measurable deterministic process; (2) an estimation problem to determine $\hat{\boldsymbol{x}}$ from the available data about the process.

Note that in this section no assumptions were made about the probability laws for the $\boldsymbol{i}_n$ and $\boldsymbol{x}_n$, and only the equation for evolution of the state was assumed to be linear.

### *9.3.3 Review of State Estimation*

The calculation of $\hat{\boldsymbol{x}}$, the conditional mathematical expectation of the state, is linked to filtering and to prediction. The appendix contains a brief discussion of these problems, and a summary of the main results.

In the linear case, the estimate $\hat{\boldsymbol{x}}$ satisfies a system of recursion relations. If the process is

$$\boldsymbol{x}_{n+1} = \mathbf{F}_n\boldsymbol{x}_n + \mathbf{H}_n\boldsymbol{u}_n + \boldsymbol{\Delta}_n, \tag{9.37}$$

$$\boldsymbol{i}_n = \mathbf{G}_n\boldsymbol{x}_n + \boldsymbol{\Delta}_n', \tag{9.38}$$

with

$$\mathrm{E}[\boldsymbol{\Delta}\boldsymbol{\Delta}^{\mathrm{T}}] = \boldsymbol{\Phi}, \qquad \mathrm{E}[\boldsymbol{\Delta}'\boldsymbol{\Delta}'^{\mathrm{T}}] = \boldsymbol{\Phi}' \tag{9.39}$$

$$\mathrm{E}[\boldsymbol{\Delta}'\boldsymbol{\Delta}^{\mathrm{T}}] = \mathbf{0},$$

then $\hat{\boldsymbol{x}}_n$ can be calculated from

$$\hat{\boldsymbol{x}}_n = \mathbf{F}_{n-1}\hat{\boldsymbol{x}}_{n-1} + \mathbf{H}_{n-1}\boldsymbol{u}_{n-1} + \mathbf{M}_n[\mathbf{G}_n(\mathbf{F}_{n-1}\hat{\boldsymbol{x}}_{n-1} + \mathbf{H}_{n-1}\boldsymbol{u}_{n-1}) - \boldsymbol{i}_n], \tag{9.40}$$

where the matrices $\mathbf{M}_n$ are generated by the matrix recurrence relations

$$\mathbf{M}_n = \boldsymbol{\Psi}_n \mathbf{G}_n^{\mathrm{T}}[\mathbf{G}_n \boldsymbol{\Psi}_n \mathbf{G}_n^{\mathrm{T}} + \boldsymbol{\Phi}_n']^{-1}, \tag{9.41}$$

$$\boldsymbol{\Psi}_{n+1} = \mathbf{F}_n[1 - \mathbf{M}_n \mathbf{G}_n]\boldsymbol{\Psi}_n[1 - \mathbf{M}_n \mathbf{G}_n]^{\mathrm{T}}\mathbf{F}_n^{\mathrm{T}} + \boldsymbol{\Phi}_n + \mathbf{F}_n \mathbf{M}_n \boldsymbol{\Phi}_n' \mathbf{M}_n^{\mathrm{T}} \mathbf{F}_n^{\mathrm{T}}, \tag{9.42}$$

with $\boldsymbol{\Psi}_n$ the covariance matrix of

$$\boldsymbol{\varepsilon}_n = \boldsymbol{x}_n - \hat{\boldsymbol{x}}_n.$$

In the recursion for $\hat{\boldsymbol{x}}_n$, the first two terms are the same as in the deterministic process equation. The term in brackets is the difference between the data predicted from the past measurements, and the data actually measured at time $n$. As modified by $\mathbf{M}_n$, it is this difference which affects the evolution of the "process model" given by the first two terms.

### 9.3.4 *Structure of the Numerical Controller*

The above results lead to the control scheme indicated in Fig. 9.1. The control system has two parts, a dynamic part, the estimator of $\hat{\boldsymbol{x}}$, and a static part, the control function $\mathbf{L}_n$. Even if the process is stationary, $\mathbf{L}_n$ and $\mathbf{M}_n$ are not constant in the case that $N$ is finite. In the case that $N \to \infty$, and the problem still retains its sense, $\mathbf{L}$ and $\mathbf{M}$ are constant, and can be obtained as the limits generated by the recursions for $\mathbf{L}_n$ and $\mathbf{M}_n$.

### 9.3.5 *Extension to the Case of Nonperiodic Measurements*

Let us consider the case in which, at the instant the control $\boldsymbol{u}_n$ is to be applied, certain information about the process is available, but obtained at previous times which are not necessarily synchronized with the times control was applied. This situation can be treated by denoting the totality of available information, whatever it may be, as $\boldsymbol{I}$, and optimizing the expected return conditionally on $\boldsymbol{I}$. The results of Sections 9.3.1 and 9.3.2 remain valid whatever the nature of $\boldsymbol{I}$, so that the separation theorem also holds in this more general case.

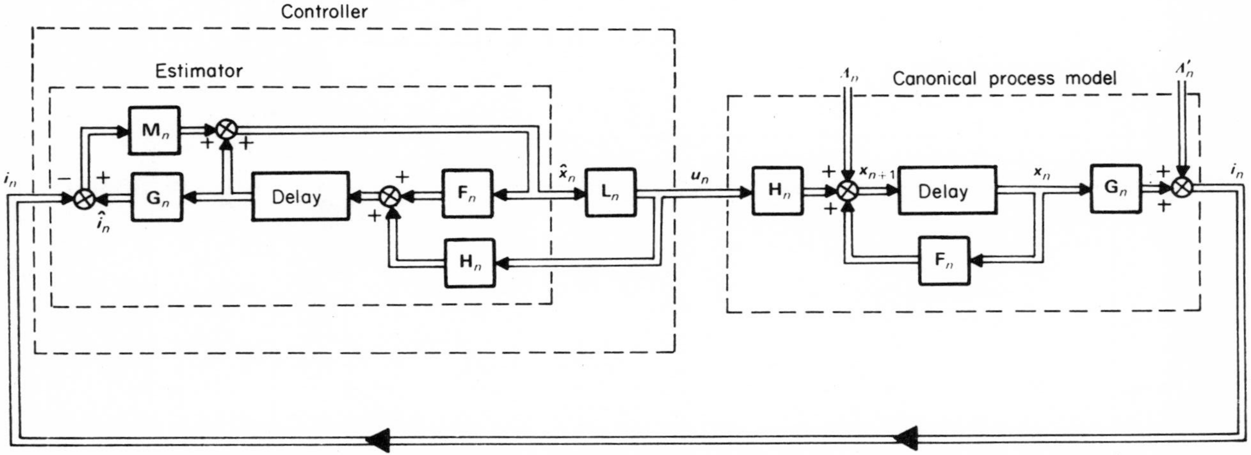


FIG. 9.1. A stochastic process and its controller.

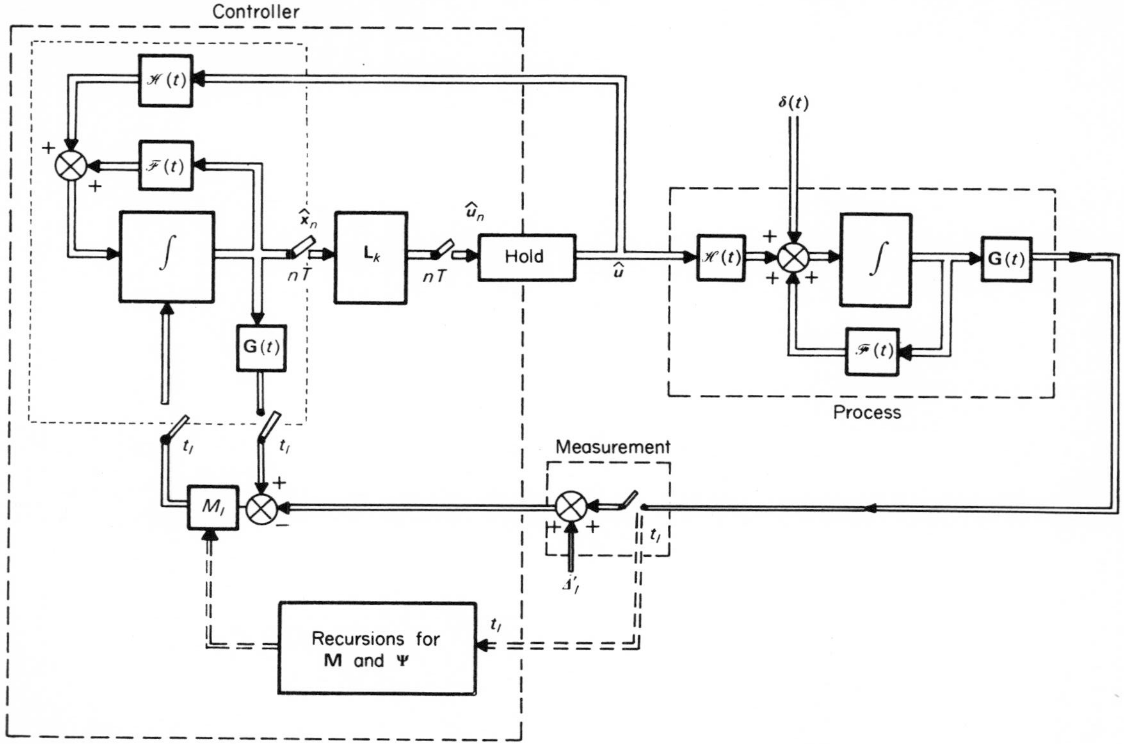


FIG. 9.2. Control of a continuous system with asynchronous sampling.

However, use of the separation theorem now requires calculation of the mathematical expectation of the state, conditioned by $\boldsymbol{I}$. This poses no particular theoretical problem, since (9.40), (9.41), and (9.42) are valid whatever the time intervals $T_i$ separating the times of arrival of the data. Specifically, consider the continuous process described by

$$\dot{\boldsymbol{x}}(t) = \mathscr{F}(t)\boldsymbol{x} + \mathscr{H}(t)\boldsymbol{u} + \boldsymbol{\delta}(t), \qquad (9.43)$$

and let $\hat{\boldsymbol{x}}_l(t_l)$ be the estimate of $\boldsymbol{x}(t)$ at the time $t_l$ of measuring the data

$$\boldsymbol{i}_l = \mathbf{G}(t_l)\boldsymbol{x}(t_l) + \boldsymbol{\Delta}_l'. \qquad (9.44)$$

For all $t$ less than $t_{l+1}$, the best estimate will be the result $\hat{\boldsymbol{x}}_l(t)$ of integrating the differential system (9.43) from time $t_l$ until time $t$, using as initial condition $\hat{\boldsymbol{x}}_l(t_l)$, letting $\boldsymbol{\delta} = \boldsymbol{0}$, and taking the optimal control $\hat{\boldsymbol{u}}(t)$ for $\boldsymbol{u}$.

At the instant $t_{l+1}$ of arrival of the next measurement, the estimate should be modified in accordance with (9.40):

$$\hat{\boldsymbol{x}}_{l+1}(t_{l+1}) = \hat{\boldsymbol{x}}_l(t_{l+1}) + \mathbf{M}_{l+1}[\mathbf{G}(t_{l+1})\hat{\boldsymbol{x}}_l(t_{l+1}) - \boldsymbol{i}_{l+1}].$$

The matrices $\mathbf{M}_l$ are obtained using a system identical to (9.41) and (9.42), but with $\mathbf{F}_l$ the value $\mathbf{A}(t_l)$ obtained by integrating the system

$$\dot{\mathbf{A}}(t) = \mathscr{F}(t)\mathbf{A}(t),$$
$$\mathbf{A}(t_{l-1}) = \mathbf{1},$$

with $\boldsymbol{\Phi}_l$ the covariance matrix of the state $\boldsymbol{x}_n$ resulting from the application of "white" noise $\boldsymbol{\delta}(t)$ from time $t_{l-1}$ till $t_l$ and with $\boldsymbol{\Phi}_l'$ the covariance matrix of the measurement noise $\boldsymbol{\Delta}_l'$. These matrices can be calculated using the methods discussed by Boudarel *et al.* [1, Vol. 1, Sect. 14.4].

All these quantities can be calculated in real time. This leads to the scheme indicated in Fig. 9.2.

**Remark.** In case the controller may possibly miss a control action, the separation theorem still remains valid. Only the last $\mathbf{L}_k$ must be changed, to conform to the results of Section 9.2.

## 9.4 Conclusions

In the general case, solution of the problem posed in Chapter 7 involves the following calculations:

a. System considerations: (1) From the information about the problem, calculate the mathematical models for the input signals, the process, and the perturbations; (2) from the system objective, determine an appropriate cost criterion; (3) using the results of (1) and (2), determine the canonical model, defined by the matrices $\mathbf{F}_n$, $\mathbf{H}_n$, $\mathbf{G}_n$, $\mathbf{\Phi}_n$, $\mathbf{\Phi}_n'$, $\mathbf{R}_1$, $\mathbf{R}_2$, and $\mathbf{R}_3$.

b. Calculate the control functions $\mathbf{L}_n$.

c. Calculate the estimator matrices $\mathbf{M}_n$.

d. Program the controller described by the recursions

$$\hat{\boldsymbol{x}}_n = \mathbf{F}_{n-1}\hat{\boldsymbol{x}}_{n-1} + \mathbf{H}_{n-1}\hat{\boldsymbol{u}}_{n-1} + \mathbf{M}_n[\mathbf{G}_n\mathbf{F}_{n-1}\hat{\boldsymbol{x}}_{n-1} - \boldsymbol{i}_n],$$

$$\hat{\boldsymbol{u}}_n = -\mathbf{L}_n\hat{\boldsymbol{x}}_n.$$

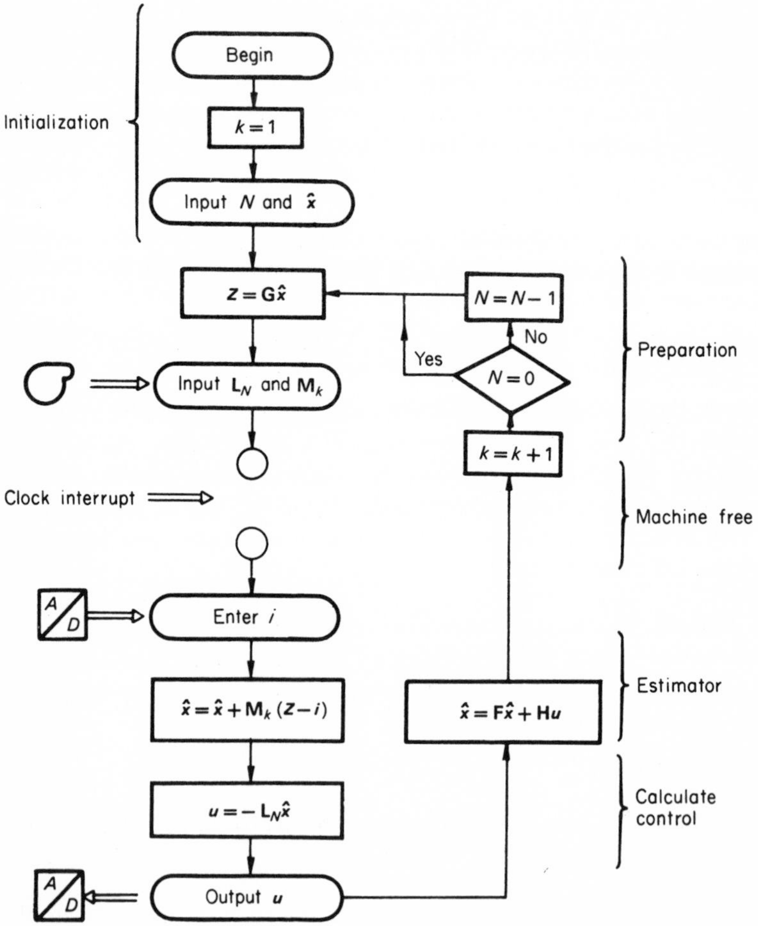


FIG. 9.3. Flow chart of the controller of a process that is not completely measurable.

The last phase in the above list corresponds to the flow chart in Fig. 9.3. The calculations have been arranged so as to reduce as much as possible the amount of computation required, and particularly the computations required between the time of input of $\boldsymbol{i}_n$ and the time of availability of the resulting $\hat{\boldsymbol{u}}_n$.

Using standard library programs, it is possible to formulate the vector equation for the discrete states, and to calculate recursively the matrices **Q**, **L**, **Ψ**, and **M**. This simplifies the synthesis problem to the point that controller synthesis becomes a simple routine task, once the process has been identified and the problem well defined. However, it is just these preliminary steps which are the most delicate, and which require experience and application of the art of engineering.

It might be remarked in closing that the calculations involved, both in the synthesis process and in the controller operation, never require computers having performance which is at all exceptional, contrary to the general situation in the case of nonlinear processes. With computers of only modest speed and memory capacity, relations (9.9) and (9.42) can be solved in a reasonable time, even for problems in which the total state vector is of fairly large dimension. This is due mainly to the repetitive nature of the calculations, and to the possibility of programming universal algorithms, which allow the use of stored-program electronic calculators.

# PART 4 | Continuous Processes

# Chapter 10 | Continuous Deterministic Processes

In this chapter we shall use dynamic programming to solve the continuous optimal control problem. We shall first indicate how a continuous process can be considered as the limit of a discrete process, for which the preceding chapters furnish the optimal solution. Using this approach, in the second section we shall obtain the optimality condition in the form of a functional equation involving the first partial derivatives of the process. In the third section some special cases are examined for which analog solutions can be obtained.

Using these results, in the fourth section we shall establish some classical theorems of the calculus of variations.

In the fifth section an approach analogous to the method of characteristics is used to establish the maximum principle. However, we shall need to assume that the second partial derivatives of the optimal return function exist, which limits the validity of our proof.

Some practical applications of the results obtained in the chapter are discussed in the last section.

## 10.1 A Continuous Process as the Limit of a Discrete Process

For some given functions $r(\boldsymbol{x}, \boldsymbol{u}, t)$ and $\boldsymbol{f}(\boldsymbol{x}, \boldsymbol{u}, t)$, each value of the parameter $\Delta$ determines a discrete process defined by

$$\begin{aligned} \boldsymbol{y}((n+1)\Delta) &= \boldsymbol{y}(n\Delta) + \Delta \cdot \boldsymbol{f}(\boldsymbol{y}(n\Delta), \boldsymbol{v}(n\Delta), n\Delta), \\ R^{\Delta} &= \sum_{n=1}^{T/\Delta} \Delta \cdot r(\boldsymbol{y}(n\Delta), \boldsymbol{v}(n\Delta), n\Delta). \end{aligned} \tag{10.1}$$

Then to each given continuous function $\boldsymbol{u}(t)$ there corresponds a solution $\boldsymbol{y}(n\Delta)$ calculated using (10.1), provided we choose

$$\boldsymbol{v}(n\Delta) = \boldsymbol{u}(n\Delta). \tag{10.2}$$

Provided $\boldsymbol{f}(\ )$ satisfies Lipschitz conditions, it is true that

$$\lim_{\Delta \to 0} \boldsymbol{y}(n\Delta) \to \boldsymbol{x}(n\Delta), \tag{10.3}$$

where $\boldsymbol{x}(t)$ is a solution of the differential system

$$\dot{\boldsymbol{x}} = \boldsymbol{f}(\boldsymbol{x}, \boldsymbol{u}, t) \tag{10.4}$$

which is the limit of the discrete system (10.1), written in the form

$$[\boldsymbol{y}((n+1)\Delta) - \boldsymbol{y}(n\Delta)]/\Delta = \boldsymbol{f}(\boldsymbol{y}(n\Delta), \boldsymbol{v}(n\Delta), n\Delta). \tag{10.5}$$

In addition, $R^{\Delta}$ tends toward the value

$$\mathscr{R} = \int_0^T r(\boldsymbol{x}, \boldsymbol{u}, t)\, \mathrm{d}t, \tag{10.6}$$

the limit of the sum being an integral.

As a result, in order to study continuous processes such as described by (10.4) and (10.6), we shall first replace them by the corresponding discrete processes of the form (10.1), to which the results developed in the earlier chapters apply, and then, in the solution to the discrete problem, let $\Delta$ tend to zero to obtain the functional optimality equations for the continuous case.

## 10.2 Establishment of the Functional Optimality Equations

### *10.2.1 Process with Bounded Horizon*

Let us seek to optimize the total return for a continuous process of the type (10.4), with $T$ fixed in advance and with a control constraint of the type

$$\boldsymbol{u}(t) \in \Omega(\boldsymbol{x}, t). \tag{10.7}$$

Let $\hat{\mathscr{R}}(\boldsymbol{x}, t)$ be the optimal return corresponding to the initial state $\boldsymbol{x}$ taken at time $t$. Since the horizon is bounded, if $\boldsymbol{x}(t)$ is unconstrained,

$$\hat{\mathscr{R}}(\boldsymbol{x}, T) \equiv 0. \tag{10.8}$$

The discrete process (10.1) corresponds to the optimality equation

$$\hat{R}^{\Delta}(y(n\Delta), n\Delta) = \underset{v(n\Delta) \in \Omega(y, n\Delta)}{\text{opt}} [\Delta \cdot r(y(n\Delta), v(n\Delta), n\Delta) + \hat{R}^{\Delta}(y((n+1)\Delta), (n+1)\Delta)]. \quad (10.9)$$

Using (10.1), it follows that

$$\hat{R}^{\Delta}(y((n+1)\Delta), (n+1)\Delta) = \hat{R}^{\Delta}(y(n\Delta) + \Delta \cdot f(y(n\Delta), v(n\Delta), n\Delta), (n+1)\Delta). \quad (10.10)$$

If $\Delta$ is small, and if it is assumed that $\hat{R}^{\Delta}$ ( ) is continuous in $y$, this expression can be expanded to first order as

$$\hat{R}^{\Delta}(y((n+1)\Delta), (n+1)\Delta) \simeq \hat{R}^{\Delta}(y(n\Delta), (n+1)\Delta) + \hat{R}_y^{\mathrm{T}}(y(n\Delta), (n+1)\Delta) f(y(n\Delta), v(n\Delta), n\Delta)\Delta, \quad (10.11)$$

where $\hat{R}_y$ is the gradient vector of $\hat{R}^{\Delta}$ with respect to $y$:

$$(\hat{R}_y)_i = \partial \hat{R}^{\Delta} / \partial y_i .$$

Substituting (10.11) into (10.9) yields

$$\hat{R}^{\Delta}(y(n\Delta), n\Delta) \simeq \underset{v \in \Omega}{\text{opt}}[\Delta \cdot r(y(n\Delta), v(n\Delta), n\Delta) + \hat{R}^{\Delta}(y(n\Delta), (n+1)\Delta) + \Delta \cdot \hat{R}_y^{\mathrm{T}}(y(n\Delta), (n+1)\Delta) f(y(n\Delta), v(n\Delta), n\Delta)] \quad (10.12)$$

or

$$[\hat{R}^{\Delta}(y(n\Delta), n\Delta) - \hat{R}^{\Delta}(y(n\Delta), (n+1)\Delta)]/\Delta \simeq \underset{v \in \Omega}{\text{opt}}[r(y(n\Delta), v(n\Delta), n\Delta) + \hat{R}_y^{\mathrm{T}}(y(n\Delta), (n+1)\Delta) f(y(n\Delta), v(n\Delta), n\Delta)]. \quad (10.13)$$

Letting $\Delta$ tend to zero in this results in the following functional equation of optimality:

$$\partial \hat{\mathscr{R}}(x, t)/\partial t = -\underset{u \in \Omega(x, t)}{\text{opt}}[r(x, u, t) + \hat{\mathscr{R}}_x^{\mathrm{T}}(x, t) f(x, u, t)]. \quad (10.14)$$

This is a partial differential equation for the optimal return. The boundary condition, which must be satisfied, is

$$\hat{\mathscr{R}}(x, T) \equiv 0 \qquad \text{for} \quad x \in \mathscr{D}, \quad (10.15)$$

where $\mathscr{D}$ is the domain of final states. Equation (10.14) will be established again by Boudarel *et al.* [1, Vol. 4]. It is known as the Hamilton-Jacobi equation, the Hamilton–Jacobi–Bellman equation, or simply as Bellman's equation.

**Remark.** Equation (10.14) can also be established directly. The return $\hat{\mathscr{R}}(\boldsymbol{x}, t)$ is equal to the return realized on the interval $[t, t + \Delta]$, plus the return corresponding to initialization at the state $\boldsymbol{x}(t + \Delta)$. The principle of optimality states that this latter return is in fact the optimal return $\hat{\mathscr{R}}(\boldsymbol{x}(t + \Delta), t + \Delta)$. Then, for small $\Delta$,

$$\hat{\mathscr{R}}(\boldsymbol{x}, t) \simeq \underset{\boldsymbol{u}(t) \in \Omega(\boldsymbol{x}, t)}{\text{opt}} [\Delta \cdot r(\boldsymbol{x}, \boldsymbol{u}, t) + \hat{\mathscr{R}}(\boldsymbol{x}(t + \Delta), t + \Delta)]. \tag{10.16}$$

Replacing $\boldsymbol{x}(t + \Delta)$ with $\boldsymbol{x}(t) + \Delta\dot{\boldsymbol{x}}(t)$, and expanding (10.16) to first order in $\Delta$, yields (10.14) upon passage to the limit as $\Delta$ tends to zero.

### 10.2.2 *Infinite Horizon*

For a stationary process, letting $T$ tend toward infinity and assuming that the problem remains sensible, results in a return $\hat{\mathscr{R}}(\boldsymbol{x}, t)$ which does not depend explicitly on $t$. Thus

$$\partial\hat{\mathscr{R}}(\boldsymbol{x}, t)/\partial t = 0,$$

and (10.14) reduces to

$$0 = \underset{\boldsymbol{u} \in \Omega(\boldsymbol{x})}{\text{opt}} [r(\boldsymbol{x}, \boldsymbol{u}) + \hat{\mathscr{R}}_x^{\mathrm{T}}(\boldsymbol{x})\boldsymbol{f}(\boldsymbol{x}, \boldsymbol{u})], \tag{10.17}$$

which is a functional equation involving only $\boldsymbol{x}$.

### 10.2.3 *Horizon Not Specified in Advance*

If the process is stationary, and if the terminal time is determined by a state condition of the form $\boldsymbol{\Phi}(\boldsymbol{x}) = \boldsymbol{0}$, the optimal return again depends only on $\boldsymbol{x}$. Thus relation (10.17) is valid, with the boundary conditions

$$\begin{aligned} \hat{\mathscr{R}}(\boldsymbol{x}) &= 0, \qquad \boldsymbol{\Phi}(\boldsymbol{x}) = \boldsymbol{0}, \\ \hat{\mathscr{R}}(\boldsymbol{x}) &\neq 0, \qquad \boldsymbol{\Phi}(\boldsymbol{x}) \neq \boldsymbol{0}. \end{aligned} \tag{10.18}$$

**Example.** Let us consider the problem of minimum-time regulation to the origin. Let $T(x)$ be the minimum time. Then

$$0 = \min_{u \in \Omega(x)} [k + T_x^{\mathrm{T}}(x) f(x, u)],$$

$$T(0) = 0,$$

$$k = 0, \qquad x = 0,$$

$$k = 1, \qquad x \neq 0.$$

## 10.3 Special Case of Unconstrained Control

When the control $u$ is free, $\Omega(x) = \mathbf{R}^p$, and if the functions $r(\ )$ and $f(\ )$ are continuous and differentiable with respect to $u$, the optimization operator in (10.14) can be eliminated. A necessary condition then results involving only partial differential equations.

### *10.3.1 The General Case*

The value of $u$ which optimizes

$$E = r(x, u, t) + \hat{\mathscr{R}}_x^{\mathrm{T}}(x, t) f(x, u, t) \tag{10.19}$$

must satisfy the first-order necessary condition

$$\partial E / \partial u_i = 0 \Rightarrow E_u = 0. \tag{10.20}$$

Thus

$$r_u(x, u, t) + \mathbf{F}_u^{\mathrm{T}}(x, u, t) \hat{\mathscr{R}}_x(x, t) = 0, \tag{10.21}$$

where $\mathbf{F}_u(x, u, t)$ is the matrix with elements $\partial f_i / \partial u_j$. Thus the optimal solution must be such that

$$\partial \hat{\mathscr{R}}(x, t) / \partial t + r(x, u, t) + \hat{\mathscr{R}}_x^{\mathrm{T}}(x, t) f(x, u, t) = 0, \tag{10.22}$$

$$r_u(x, u, t) + \mathbf{F}_u^{\mathrm{T}}(x, u, t) \hat{\mathscr{R}}_x(x, t) = 0, \tag{10.23}$$

in the case of a bounded horizon, or

$$r(x, u) + \hat{\mathscr{R}}_x^{\mathrm{T}}(x) f(x, u) = 0, \tag{10.24}$$

$$r_u(x, u) + \mathbf{F}_u^{\mathrm{T}}(x, u) \hat{\mathscr{R}}_x(x) = 0, \tag{10.25}$$

in the case of an infinite horizon, or a horizon not specified in advance.

**Remark 1.** The matrix with elements $\partial^2 E/\partial u_i u_j$ must be nonnegative definite to satisfy the second-order conditions for a minimum, and non-positive definite to satisfy the conditions for a maximum.

**Remark 2.** If it is possible to solve the equation $\boldsymbol{E_u}=\boldsymbol{0}$ for $\boldsymbol{u}$, the solution can be used to eliminate $\boldsymbol{u}$ from (10.22) to obtain a single partial differential equation involving only $\boldsymbol{x}$ and $t$.

### *10.3.2 First-Order Systems*

For a first-order stationary system with unspecified (or infinite) horizon there results the system

$$r(x, u) + \mathrm{d}[\mathscr{R}(x)]/\mathrm{d}x\, f(x, u) = 0, \tag{10.26}$$

$$\partial r(x, u)/\partial u + \mathrm{d}[\mathscr{R}(x)]/\mathrm{d}x\ \partial f(x, u)/\partial u = 0. \tag{10.27}$$

That a solution exist requires that

$$\begin{vmatrix} r(x, u) & f(x, u) \\ \partial r(x, u)/\partial u & \partial f(x, u)/\partial u \end{vmatrix} = 0,$$

or

$$r(x, u)\ \partial f(x, u)/\partial u - f(x, u)\ \partial r(x, u)/\partial u = 0. \tag{10.28}$$

Relation (10.28) must be satisfied by the optimizing $u(x)$. Thus the sought control must be one of the possible solutions to (10.28). To determine which solution is the one of interest, the various $\mathscr{R}(x)$ can be calculated, and those sought which satisfy the optimality equation.

**Example.** Let the process be linear,

$$\dot{x} = x + u,$$

and let us seek the regulator control which minimizes

$$\mathscr{R} = \int_0^\infty (x^2 + u^2)\,\mathrm{d}t.$$

Thus

$$r(\,) = x^2 + u^2 \rightarrow \partial r/\partial u = 2u,$$

$$f(\,) = x + u \rightarrow \partial f/\partial u = 1,$$

so that the implicit relation (10.28) becomes

$$x^2 + u^2 - 2u(x + u) = 0.$$

There are two solutions,

$$u = -(1 \pm \sqrt{2})x.$$

Relation (10.26) then yields

$$\mathrm{d}\hat{\mathscr{R}}/\mathrm{d}x = -(\partial r/\partial u)/(\partial f/\partial u) = -2u = 2(1 \pm \sqrt{2})x,$$

which, when integrated, results in

$$\hat{\mathscr{R}}(x) = A + (1 \pm \sqrt{2})x^2.$$

The condition $\hat{\mathscr{R}}(0) = 0$ allows the constant to be evaluated:

$$A = 0.$$

Since $\hat{\mathscr{R}}(x)$ is the integral of a function, $x^2 + u^2$, which is always positive, it is itself positive, and the plus sign must be chosen above. Thus

$$\hat{u} = -(1 + \sqrt{2})x, \qquad \hat{\mathscr{R}}(x) = (1 + \sqrt{2})x^2.$$

The closed-loop system using the optimal control satisfies the differential equation

$$\dot{x} = x - (1 + \sqrt{2})x = -\sqrt{2}x,$$

so that

$$x(t) = X_0 \mathrm{e}^{-\sqrt{2}t},$$

and hence

$$u(t) = -(1 + \sqrt{2})X_0\,\mathrm{e}^{-\sqrt{2}t}.$$

The final time is $T = \infty$, and thus

$$\hat{\mathscr{R}}(X_0) = \mathrm{X}_0{}^2 \int_0^\infty [\mathrm{e}^{-2\sqrt{2}t} + (1 + \sqrt{2})^2 \mathrm{e}^{-2\sqrt{2}t}]\,\mathrm{d}t = (1 + \sqrt{2})X_0{}^2.$$

### *10.3.3 Linear System with Quadratic Criterion*

Let the process be linear,

$$\dot{\boldsymbol{x}} = \mathbf{F}\boldsymbol{x} + \mathbf{H}\boldsymbol{u}, \qquad (10.29)$$

and let the cost criterion be quadratic:

$$r(\boldsymbol{x}, \boldsymbol{u}) = \boldsymbol{x}^{\mathrm{T}}\mathbf{A}\boldsymbol{x} + 2\boldsymbol{x}^{\mathrm{T}}\mathbf{B}\boldsymbol{u} + \boldsymbol{u}^{\mathrm{T}}\mathbf{C}\boldsymbol{u}. \tag{10.30}$$

Condition (10.21) for this case becomes

$$2\mathbf{B}^{\mathrm{T}}\boldsymbol{x} + 2\mathbf{C}\boldsymbol{u} + \mathbf{H}^{\mathrm{T}}\hat{\mathscr{R}}_x = \boldsymbol{0}, \tag{10.31}$$

so that

$$\hat{\boldsymbol{u}} = -\mathbf{C}^{-1}[\mathbf{B}^{\mathrm{T}}\boldsymbol{x} + \tfrac{1}{2}\mathbf{H}^{\mathrm{T}}\hat{\mathscr{R}}_x]. \tag{10.32}$$

The return is quadratic, of the form

$$\hat{\mathscr{R}}(\boldsymbol{x}, t) = \boldsymbol{x}^{\mathrm{T}}\mathbf{Q}(t)\boldsymbol{x}. \tag{10.33}$$

Thus

$$\hat{\mathscr{R}}_t = \boldsymbol{x}^{\mathrm{T}}\dot{\mathbf{Q}}\boldsymbol{x},$$
$$\hat{\mathscr{R}}_x = 2\mathbf{Q}\boldsymbol{x},$$

from which

$$\hat{\boldsymbol{u}} = -\mathbf{C}^{-1}[\mathbf{B} + \mathbf{Q}\mathbf{H}]^{\mathrm{T}}\boldsymbol{x}. \tag{10.34}$$

In this case it is necessary that $\mathbf{C}^{-1}$ exist. Otherwise the optimization problem has no meaning.

The optimal control is linear. The matrix $\mathbf{Q}(t)$ can be found by substituting $\boldsymbol{u}$ into the optimality equation. The result is

$$\dot{\mathbf{Q}} = -\mathbf{A} - \mathbf{F}^{\mathrm{T}}\mathbf{Q} - \mathbf{Q}\mathbf{F} + [\mathbf{B} + \mathbf{Q}\mathbf{H}]\mathbf{C}^{-1}[\mathbf{B} + \mathbf{Q}\mathbf{H}]^{\mathrm{T}}, \tag{10.35}$$

which is a nonlinear matrix differential equation of the Riccati type. In general, it can not be solved analytically, but must be integrated numerically, or by analog computation.

In the case that the final state is free and the time $T$ fixed, there results

$$\hat{\mathscr{R}}(\boldsymbol{x}, T) = 0, \qquad \forall \boldsymbol{x},$$

which requires that

$$\mathbf{Q}(T) = \mathbf{0}. \tag{10.36}$$

Equation (10.35) should thus be integrated backward, using the "initial" value (10.36).

When $T \to \infty$, $\dot{\mathbf{Q}} \to 0$ and $\mathbf{Q}(\infty)$ then satisfies a nonlinear matrix algebraic equation. This is analogous to the results found in Section 5.6.

## 10.4 Application to the Calculus of Variations

We shall apply the functional equation of optimality, (10.14), to the following problem from the calculus of variations: Find the function $x(t)$ which minimizes

$$\int_0^T r(x, \dot{x}, t)\, \mathrm{d}t, \tag{10.37}$$

where $r$ is continuously differentiable in both $x$ and $\dot{x}$. We shall be able to establish very easily the Euler equation, and the Legendre condition.

This problem corresponds to the case in which

$$\dot{x} = u,$$

and $u$ is unconstrained. Thus, from (10.14),

$$-\partial\hat{\mathscr{R}}(x, t)/\partial t = \underset{u}{\mathrm{opt}}[r(x, u, t) + u\, \partial\hat{\mathscr{R}}(x, t)/\partial x]. \tag{10.38}$$

Differentiating the term in brackets with respect to $u$ leads to

$$\partial r/\partial u + \partial\hat{\mathscr{R}}/\partial x = 0, \tag{10.39}$$

which must be satisfied by $u$. Considering $u$ in (10.38) to have the value $\hat{u}$, (10.38) yields a second relation for $u$:

$$\partial\hat{\mathscr{R}}/\partial t + r + \partial\hat{\mathscr{R}}/\partial x \; u = 0. \tag{10.40}$$

Differentiating (10.39) with respect to $t$ and (10.40) with respect to $x$ results in

$$\frac{\mathrm{d}}{\mathrm{d}t}\frac{\partial r}{\partial u} + \frac{\partial^2\hat{\mathscr{R}}}{\partial x\, \partial t} + \frac{\partial^2\hat{\mathscr{R}}}{\partial x^2}\, \dot{x} = 0,$$

$$\frac{\partial^2\hat{\mathscr{R}}}{\partial x\, \partial t} + \frac{\partial r}{\partial x} + \frac{\partial^2\hat{\mathscr{R}}}{\partial x^2}\, u = 0.$$

Subtracting these, and taking account that $u = \dot{x}$, yields

$$\frac{\mathrm{d}}{\mathrm{d}t}\left[\frac{\partial r(x, \dot{x}, t)}{\partial \dot{x}}\right] - \frac{\partial r(x, \dot{x}, t)}{\partial x} = 0, \tag{10.41}$$

which is precisely the Euler equation.

We have thus far used only the first-order necessary condition that $u$ be optimum. In the case of minimization, $u$ should also be such that

$$\partial^2[\;]/\partial u^2 \geq 0,$$

or, in this case,

$$\partial^2 r/\partial u^2 \geq 0,$$

which is to say

$$\partial^2 r(x, \dot{x}, t)/\partial \dot{x}^2 \geq 0, \tag{10.42}$$

which is the Legendre condition.

These results can be easily generalized to the vector case, to obtain the Euler equation,

$$\mathrm{d}[\boldsymbol{r}_{\dot{x}}(\boldsymbol{x}, \dot{\boldsymbol{x}}, t)]/\mathrm{d}t - \boldsymbol{r}_x(\boldsymbol{x}, \dot{\boldsymbol{x}}, t) = 0, \tag{10.43}$$

and the Legendre condition that the matrix with elements

$$\partial^2 r(\boldsymbol{x}, \dot{\boldsymbol{x}}, t)/\partial \dot{x}_i \, \partial \dot{x}_j \tag{10.44}$$

be nonnegative definite.

## 10.5 The Maximum Principle

We have seen that, if the control is unconstrained, the optimization operator in the Hamilton–Jacobi equation can be removed from consideration. This led to the Euler equation, which is a necessary condition for optimality of the solution. This second-order differential equation can be replaced by a pair of first-order equations, the canonical equations of Hamilton. In the more general case, in which the control must satisfy a constraint of the form $\boldsymbol{u} \in \Omega(\boldsymbol{x})$, it is also possible to replace the Hamilton–Jacobi equation by a first-order differential system. However, the optimization operator can no longer be suppressed. The necessary condition which results from this procedure is the maximum principle of Pontryagin.

### *10.5.1 The Case of an Infinite or Unspecified Horizon*

If the horizon is either infinite, or not specified in advance, the optimality equation is

$$0 = \underset{\boldsymbol{u} \in \Omega(\boldsymbol{x})}{\mathrm{opt}} \, [r(\boldsymbol{x}, \boldsymbol{u}) + \hat{\mathscr{R}}_x^{\mathrm{T}}(\boldsymbol{x})\boldsymbol{f}(\boldsymbol{x}, \boldsymbol{u})]. \tag{10.45}$$

For some given initial state, let $\hat{\boldsymbol{u}}(t)$ be the optimal control, and let $\hat{\boldsymbol{x}}(t)$ be the corresponding state trajectory. Then using these in (10.45) yields

$$0 = r(\hat{\boldsymbol{x}}(t), \hat{\boldsymbol{u}}(t)) + \hat{\mathscr{R}}_x^{\mathrm{T}}(\hat{\boldsymbol{x}}(t))\boldsymbol{f}(\hat{\boldsymbol{x}}(t), \hat{\boldsymbol{u}}(t)). \tag{10.46}$$

Now let

$$\boldsymbol{\Psi}(t) = \hat{\mathscr{R}}_x(\hat{\boldsymbol{x}}(t)), \tag{10.47}$$

and define

$$H(\boldsymbol{u}) = r(\boldsymbol{x}, \boldsymbol{u}) + \boldsymbol{\Psi}^{\mathrm{T}}\dot{\boldsymbol{x}}. \tag{10.48}$$

This function $H$ is the Hamiltonian for the problem. The optimal control $\hat{\boldsymbol{u}}$ is the value of $\boldsymbol{u} \in \Omega(\boldsymbol{x})$ which optimizes the Hamiltonian, since, from (10.45),

$$0 = \hat{H}(\hat{\boldsymbol{u}}) = \underset{\boldsymbol{u}\in\Omega}{\mathrm{opt}}\, H(\boldsymbol{u}). \tag{10.49}$$

From (10.47),

$$\begin{aligned} \mathrm{d}\boldsymbol{\Psi}/\mathrm{d}t &= \mathrm{d}[\hat{\mathscr{R}}_x(\boldsymbol{x})]/\mathrm{d}t \\ &= \hat{\mathfrak{R}}_{xx}\dot{\boldsymbol{x}} = \hat{\mathfrak{R}}_{xx}\boldsymbol{f}(\boldsymbol{x}, \boldsymbol{u}). \end{aligned} \tag{10.50}$$

Differentiating (10.46) partially with respect to $\boldsymbol{x}$,

$$\boldsymbol{0} = \boldsymbol{r}_x(\boldsymbol{x}, \boldsymbol{u}) + \hat{\mathfrak{R}}_{xx}\boldsymbol{f}(\boldsymbol{x}, \boldsymbol{u}) + \mathbf{F}_x^{\mathrm{T}}(\boldsymbol{x}, \boldsymbol{u})\hat{\mathscr{R}}_x, \tag{10.51}$$

where $\boldsymbol{r}_x(\boldsymbol{x}, \boldsymbol{u})$ is the vector with elements $\partial r/\partial x_i$, and $\mathbf{F}_x(\boldsymbol{x}, \boldsymbol{u})$ is the matrix with elements $\partial f_i/\partial x_j$. Combining relations (10.50) and (10.51) leads to

$$\dot{\boldsymbol{\Psi}} = -\mathbf{F}_x^{\mathrm{T}}(\boldsymbol{x}, \boldsymbol{u})\boldsymbol{\Psi} - \boldsymbol{r}_x(\boldsymbol{x}, \boldsymbol{u}). \tag{10.52}$$

We thus have the following necessary condition, which is the maximum principle: If $\{\hat{\boldsymbol{u}}(t), \hat{\boldsymbol{x}}(t)\}$ is an optimal solution to the problem, then there exists a vector $\boldsymbol{\Psi}(t)$ such that

$$\mathrm{d}\boldsymbol{\Psi}(t)/\mathrm{d}t = -\mathbf{F}_x^{\mathrm{T}}(\hat{\boldsymbol{x}}(t), \hat{\boldsymbol{u}}(t))\boldsymbol{\Psi}(t) - \boldsymbol{r}_x(\hat{\boldsymbol{x}}(t), \hat{\boldsymbol{u}}(t)), \tag{10.53}$$

$$\begin{aligned} 0 &= r(\hat{\boldsymbol{x}}(t), \hat{\boldsymbol{u}}(t)) + \boldsymbol{\Psi}^{\mathrm{T}}(t)\boldsymbol{f}(\hat{\boldsymbol{x}}(t), \hat{\boldsymbol{u}}(t)) \\ &= \underset{\boldsymbol{u}\in\Omega(\boldsymbol{x}(t))}{\mathrm{opt}} [r(\hat{\boldsymbol{x}}(t), \boldsymbol{u}) + \boldsymbol{\Psi}^{\mathrm{T}}(t)\boldsymbol{f}(\hat{\boldsymbol{x}}(t), \boldsymbol{u})]. \end{aligned} \tag{10.54}$$

**Remark.** For problems in which the horizon is determined by stopping conditions of the form $\varphi(\boldsymbol{x}) = \boldsymbol{0}$, there is a boundary condition that $\hat{\mathscr{R}}(\boldsymbol{x}) \equiv 0$ for $\varphi(\boldsymbol{x}) = \boldsymbol{0}$. This requires that every variation $\delta\boldsymbol{x}$ of the final state $\boldsymbol{x}_{\mathrm{f}}$ which is compatible with the terminal constraint lead to a zero variation of $\hat{\mathscr{R}}(\boldsymbol{x})$. Thus $\hat{\mathscr{R}}_x = \boldsymbol{\Psi}_{\mathrm{f}}$ must be normal to the plane tangent to the hypersurface defined by the terminal constraint at the point $\boldsymbol{x}_{\mathrm{f}}$. This is the transversality condition.

### 10.5.2 *The Case of a Finite Horizon*

Proceeding as above, let

$$\boldsymbol{\Psi}(t) = \hat{\mathscr{R}}_x(\boldsymbol{x}(t), t),$$

so that

$$\mathrm{d}\boldsymbol{\Psi}(t)/dt = \hat{\mathfrak{R}}_{xx}\dot{\boldsymbol{x}} + \boldsymbol{\Psi}_t.$$

For $\boldsymbol{u} = \hat{\boldsymbol{u}}(t)$, the optimality equation becomes

$$\partial\hat{\mathscr{R}}/\partial t = -[r(\hat{\boldsymbol{x}}, \hat{\boldsymbol{u}}) + \boldsymbol{\Psi}^{\mathrm{T}}\boldsymbol{f}(\hat{\boldsymbol{x}}, \hat{\boldsymbol{u}})] = -\hat{H}.$$

Differentiating this partially with respect to $\boldsymbol{x}$ yields

$$\hat{\mathscr{R}}_{xt} = -\boldsymbol{r}_x(\boldsymbol{x}, \boldsymbol{u}) - \mathbf{F}_x^{\mathrm{T}}(\boldsymbol{x}, \boldsymbol{u})\hat{\mathscr{R}}_x - \hat{\mathfrak{R}}_{xx}\boldsymbol{f}(\boldsymbol{x}, \boldsymbol{u}),$$

so that

$$\dot{\boldsymbol{\Psi}} = -\mathbf{F}_x^{\mathrm{T}}(\boldsymbol{x}, \boldsymbol{u})\boldsymbol{\Psi} - \boldsymbol{r}_x(\boldsymbol{x}, \boldsymbol{u}).$$

We thus obtain for $\boldsymbol{\Psi}(t)$ the same differential equation as before. The necessary condition is also the same as before, except that the Hamiltonian no longer vanishes for $\boldsymbol{u} = \hat{\boldsymbol{u}}$. Rather, it takes a value such that $\hat{H} = -(\partial\hat{\mathscr{R}}/\partial t)$. If terminal constraints are present, the transversality conditions remain valid. If the final state is free, these conditions reduce to

$$\boldsymbol{\Psi}_{\mathrm{f}} = \boldsymbol{0},$$

since

$$\hat{\mathscr{R}}(\boldsymbol{x}_{\mathrm{f}}) \equiv 0, \qquad \forall x_{\mathrm{f}} \qquad \Rightarrow \hat{\mathscr{R}}_x = \boldsymbol{0}.$$

### 10.5.3 *Discussion*

Thus we have apparently arrived easily at the maximum principle, starting from the Hamilton–Jacobi equation, the functional equation of optimality. However, the above development assumed implicitly that the return $\hat{\mathscr{R}}(\boldsymbol{x})$ was differentiable to second order in $\boldsymbol{x}$. In fact, even for simple problems, that is not always the case. As a result, the above demonstration is not general. In addition, since it is very difficult to establish *a priori* the existence of $\hat{\mathfrak{R}}_{\boldsymbol{xx}}$, it is not even possible to determine easily for what problems the above procedure would be valid. However, the continuity of $\hat{\mathscr{R}}(\boldsymbol{x})$ often allows construction of the total solution by finding the solutions in each domain for which $\hat{\mathscr{R}}_x$ is continuous.

However, our aim was not to demonstrate the maximum principle, which has already been done rigorously using a different approach, but rather to show how the intuitive ideas of dynamic programming lead to its formulation. It is, in fact, possible to establish the maximum principle rigorously from a dynamic programming formulation [31], but only by introducing technical complications necessary to define rigorously the term which we have denoted $\hat{\mathfrak{R}}_{xx}$. From the practical point of view the above plausibility argument, in spite of everything, indicates that the quantities entering into the maximum principle have physical significance. In particular, the vector $\boldsymbol{\Psi}$ represents the gradient, with respect to $\boldsymbol{x}$, of the optimal return, and the optimal value of the Hamiltonian, except for sign, is the "gradient" of the optimal return with respect to time.

Boudarel *et al.* [1, Vol. 4], obtain the same result starting with the Hamilton–Jacobi equations.

## 10.6 Practical Solution

In this chapter, we have shown that the solution to the optimal control problem can be obtained starting with a functional equation of optimality, the Hamilton–Jacobi–Bellman equation. In addition, by various transformations, we have developed from this the classical results of the calculus of variations, and in particular the Euler equation, and the Pontryagin maximum principle. We shall not elaborate on the practical difficulties involved in using these latter formulas, since they are examined in detail by Boudarel *et al.* [1, Vol. 4]. We shall only remark that, in general, these formulas lead only to the optimal control for a particular initial state, and not directly to the synthesis of an optimal controller. In addition, calculation of an optimal control requires solution of a differential system with simultaneous initial and final conditions, a so-called two-point boundary value problem. This leads to difficulties in the solution.

Three things are evident in examining the optimality equation: It involves partial derivatives, the optimal return does not enter explicitly, and when $\boldsymbol{u}$ can be eliminated by using the optimality condition a first-order partial differential system can be obtained, but one which is nonlinear. It can be concluded from this that an exact analytic solution is in general impossible.

To circumvent these difficulties arising from the presence of the partial differentials, one might consider replacing them with finite-difference approximations, or perhaps $\hat{\mathscr{R}}(\boldsymbol{x}, t)$ might be approximated by a series development, and the coefficients related using the functional equation. In fact, the most successful method seems to be simply to replace the continuous problem by a corresponding discrete problem, of which it is the limit. The problem then reverts to the discrete case, and the numerical methods discussed in Chapter 4 may be applied. The discrete case is simpler than the continuous case, in that the difficulties encountered were essentially ones of numerical order.

In conclusion, it seems that the application of dynamic programming to the solution of the continuous optimal control problem is largely of theoretical interest, since the simplest method of solution is to revert to the discrete case. In addition, all the properties of the continuous solution can be obtained as easily, and with full rigor, using the variational methods discussed by Boudarel *et al.* [1, Vol. 4].

# Chapter 11 | Continuous Stochastic Processes

In this chapter, we shall consider briefly the case of a continuous stochastic process. In order that the presentation not be too weighty, we shall limit ourselves to processes which have a completely measurable state. Even in the case of discrete processes, if the state is available only through noisy measurements, the theory becomes somewhat complicated. In the continuous case, if the state measurements are noisy, the theoretical results become quite elaborate involving functional equations with partial derivatives, and their practical utilization seems remote (except in the linear case).

We shall first study processes with continuous states, for which we shall recover, in the linear-quadratic case, results analogous to those found in the discrete-time case (a linear control law, and a separation theorem). In the final section discrete-state systems will be considered, for which simple results exist.

## 11.1 Continuous Stochastic Processes with Continuous States

### *11.1.1 Definition of the Problem*

Let us consider a process with state vector $\boldsymbol{x}(t)$, to which are applied controls $\boldsymbol{u}(t)$ chosen from some domain $\mathscr{D}$. The future evolution of the state is random, and at time $t + \Delta t$

$$\Delta \boldsymbol{x} = \boldsymbol{x}(t + \Delta t) - \boldsymbol{x}(t) \tag{11.1}$$

is a random vector.

Suppose that the process is Markov (see Boudarel *et al.* [1, Vol. 1, Chap. 14]), and that, as $\Delta t$ tends to zero, the first moments of $\mathbf{\Delta x}$ are of the form

$$E[\mathbf{\Delta x}] \simeq \boldsymbol{f}(\boldsymbol{x}, \boldsymbol{u})\,\Delta t, \tag{11.2}$$

$$E[\mathbf{\Delta x}\,\mathbf{\Delta x}^{\mathrm{T}}] \simeq \boldsymbol{\Phi}(\boldsymbol{x}, \boldsymbol{u})\,\Delta t. \tag{11.3}$$

The higher moments are assumed to be of order greater than the first in $\Delta t$.

In particular, these conditions hold when the process can be represented by a differential equation

$$\dot{\boldsymbol{x}} = \boldsymbol{f}(\boldsymbol{x}, \boldsymbol{u}) + \mathbf{G}(\boldsymbol{x}, \boldsymbol{u})\boldsymbol{z}, \tag{11.4}$$

where $\boldsymbol{z}$ is a "white" noise vector with correlation matrix

$$E[\boldsymbol{z}(t)\boldsymbol{z}^{\mathrm{T}}(t+\tau)] = \boldsymbol{\Gamma}\delta(\tau), \tag{11.5}$$

with $\delta(\tau)$ the Dirac "function." In this case,

$$\boldsymbol{\Phi}(\boldsymbol{x}, \boldsymbol{u}) = \mathbf{G}(\boldsymbol{x}, \boldsymbol{u})\boldsymbol{\Gamma}\,\mathbf{G}^{\mathrm{T}}(\boldsymbol{x}, \boldsymbol{u}). \tag{11.6}$$

### *11.1.2 Solution of the Problem*

The control law is to be found which optimizes the mathematical expectation of the total future return, up to the final time $T$.

Let $\hat{R}(\boldsymbol{x}, t)$ be the mathematical expectation of the future return, using an optimal policy, starting from the initial state $\boldsymbol{x}$ at time $t$. For small $\Delta t$, the method of dynamic programming relates $\hat{R}(\boldsymbol{x} + \mathbf{\Delta x}, t + \Delta t)$ to $\hat{R}(\boldsymbol{x}, t)$ through a functional equation obtained from the principle of optimality:

$$\hat{R}(\boldsymbol{x}, t) = \underset{\boldsymbol{u} \in \mathscr{D}}{\mathrm{opt}}\,[r(\boldsymbol{x}, \boldsymbol{u})\,\Delta t + E\{\hat{R}(\boldsymbol{x} + \mathbf{\Delta x}, t + \Delta t)\}]. \tag{11.7}$$

The expectation in (11.7) is conditional on knowledge of $\boldsymbol{x}$, and $\boldsymbol{x} + \mathbf{\Delta x}$ then corresponds to the state at $t + \Delta t$ resulting from application of the control $\boldsymbol{u}$ over that interval. Expanding in a series,

$$\hat{R}(\boldsymbol{x} + \mathbf{\Delta x}, t + \Delta t) = \hat{R}(\boldsymbol{x}, t) + \hat{R}_t\,\Delta t + \hat{\boldsymbol{R}}_x^{\mathrm{T}}\,\mathbf{\Delta x} + \tfrac{1}{2}\,\mathbf{\Delta x}^{\mathrm{T}}\hat{\mathbf{R}}_{xx}\,\mathbf{\Delta x} + \cdots. \tag{11.8}$$

Because of (11.2) and (11.3),

$$\mathrm{E}[\hat{R}_x{}^{\mathrm{T}}\,\Delta x] = \hat{R}_x{}^{\mathrm{T}}\mathrm{E}[\Delta x] \simeq \hat{R}_x{}^{\mathrm{T}} f(x, u)\,\Delta t, \tag{11.9}$$

$$\begin{aligned}\mathrm{E}[\Delta x^{\mathrm{T}}\hat{\mathbf{R}}_{xx}\,\Delta x] &= \mathrm{E}[\mathrm{trace}(\hat{\mathbf{R}}_{xx}\,\Delta x\,\Delta x^{\mathrm{T}})]\\ &= \mathrm{trace}\{\hat{\mathbf{R}}_{xx}\,\mathrm{E}[\Delta x\,\Delta x^{\mathrm{T}}]\}\\ &\simeq \mathrm{trace}\{\hat{\mathbf{R}}_{xx}\,\mathbf{\Phi}\}\,\Delta t.\end{aligned} \tag{11.10}$$

Substituting (11.8) and (11.9) into (11.7), and passing to the limit as $\Delta t$ tends to zero, yields

$$\hat{R}_t + \underset{u \in \mathcal{D}}{\mathrm{opt}}\,\{r(x, u) + \hat{R}_x{}^{\mathrm{T}} f(x, u) + \tfrac{1}{2}\,\mathrm{trace}[\hat{\mathbf{R}}_{xx}\,\mathbf{\Phi}(x, u)]\} = 0, \tag{11.11}$$

with the boundary condition that $\hat{R}(x, T) = 0$ for all $x$ in the domain of terminal values.

### *11.1.3 Discussion*

a. Using the operator $L_u[\ ]$ introduced by Boudarel *et al.* [1, Vol. 1, Sect. 14.2.4], Eq. (11.11) can be written simply

$$\underset{u \in \mathcal{D}}{\mathrm{opt}}\,\{r(x, u) + L_u[\hat{R}(x, t)]\} = 0. \tag{11.12}$$

b. For $\mathbf{\Phi} \to \mathbf{0}$, the stochastic process becomes deterministic, and the functional equation obtained in the preceding chapter is recovered here.

c. So far as practical use of this theoretical result is concerned, the same remarks can be made as for the continuous deterministic case, since the presence of second partial derivatives here again complicates the situation.

## 11.2 Linear Systems with Quadratic Criteria

Let us consider the linear (not necessarily stationary) process

$$\dot{x} = \mathbf{F}x + \mathbf{H}u + \mathbf{G}z, \tag{11.13}$$

with $z$ being a white noise vector with correlation matrix $\mathbf{\Gamma}\,\delta(t)$. Let the cost criterion be quadratic:

$$r(x, u) = x^{\mathrm{T}}\mathbf{A}x + 2x^{\mathrm{T}}\mathbf{B}u + u^{\mathrm{T}}\mathbf{C}u. \tag{11.14}$$

We shall show that the functional optimality equation (11.11), in this case, has a solution of the form

$$\hat{R}(x, t) = x^{\mathrm{T}}\mathbf{Q}(t)x + k(t). \tag{11.15}$$

Assuming (11.15),

$$\hat{R}_t = \boldsymbol{x}^{\mathrm{T}}\dot{\mathbf{Q}}\boldsymbol{x} + \dot{k},$$

$$\hat{\boldsymbol{R}}_x = 2\mathbf{Q}\boldsymbol{x},$$

$$\hat{\mathbf{R}}_{xx} = 2\mathbf{Q},$$

and the term to be optimized becomes

$$\boldsymbol{x}^{\mathrm{T}}\mathbf{A}\boldsymbol{x} + 2\boldsymbol{x}^{\mathrm{T}}\mathbf{B}\boldsymbol{u} + \boldsymbol{u}^{\mathrm{T}}\mathbf{C}\boldsymbol{u} + 2\boldsymbol{x}^{\mathrm{T}}\mathbf{Q}(\mathbf{F}\boldsymbol{x} + \mathbf{H}\boldsymbol{u}) + \mathrm{trace}[\mathbf{Q}\mathbf{G}\boldsymbol{\Gamma}\mathbf{G}^{\mathrm{T}}].$$

The optimum is obtained for

$$2\mathbf{B}^{\mathrm{T}}\boldsymbol{x} + 2\mathbf{C}\hat{\boldsymbol{u}} + 2\mathbf{H}^{\mathrm{T}}\mathbf{Q}\boldsymbol{x} = \boldsymbol{0},$$

so that

$$\hat{\boldsymbol{u}} = -\mathbf{C}^{-1}[\mathbf{B} + \mathbf{Q}\mathbf{H}]^{\mathrm{T}}\boldsymbol{x}. \tag{11.16}$$

Substituting the solution (11.16) into the functional equation (11.11), and separating the terms quadratic in $\boldsymbol{x}$ from those independent of $\boldsymbol{x}$, leads to

$$\dot{\mathbf{Q}}(t) = -\mathbf{A} - \mathbf{Q}(t)\mathbf{F} - \mathbf{F}^{\mathrm{T}}\mathbf{Q}(t) + [\mathbf{B} + \mathbf{Q}(t)\mathbf{H}]\mathbf{C}^{-1}[\mathbf{B} + \mathbf{Q}(t)\mathbf{H}]^{\mathrm{T}}, \tag{11.17}$$

$$\dot{k}(t) = -\,\mathrm{trace}[\mathbf{Q}(t)\mathbf{G}\boldsymbol{\Gamma}\mathbf{G}^{\mathrm{T}}]. \tag{11.18}$$

The relation (11.17) is a matrix nonlinear differential equation of the Riccati type.

Equation (11.17) is the same as that obtained in the deterministic case, and thus the control law is unchanged. This result expresses the separation property for the control of a linear process. Relation (11.18) allows calculation of the degradation of the return, resulting from the presence of perturbations $\boldsymbol{z}$.

Note that the problem has a solution only if $\mathbf{C}^{-1}$ exists, since otherwise it is possible to find control laws which lead to zero return even for $T \to 0$. When the final state is free, $\mathbf{Q}(t_{\mathrm{f}}) = \mathbf{0}$, and the matrix Riccati equation can be integrated in reverse time.

**Remark.** It can be shown that the separation theorem still holds true, even in the case that the state is not completely measurable. This result can be obtained either directly, or by passing to the limit in the discrete case. Thus an estimator $\hat{\boldsymbol{x}}$ of $\boldsymbol{x}$ can be used in (11.16) to obtain $\hat{\boldsymbol{u}}$. The main results of the theory of state estimation for continuous linear processes are summarized in the appendix.

## 11.3 Continuous Systems with Discrete States

### *11.3.1 Definition of the Problem*

Consider a process with discrete states, denoted by $i \in [1, \ldots, m]$. Let $P_{ij}(t_0, [\boldsymbol{u}]_{t_0}^{t_1}, t_1)$ be the probability that at time $t_1$ the system is in state $j$, if at time $t_0$ it was in state $i$, and if the control $\boldsymbol{u}(t)$ was applied over the interval from $t_0$ to $t_1$. Suppose further that for $t_1 \to t_0$ this probability behaves as

$$P_{ij}(t, \boldsymbol{u}, t + \Delta t) \to \begin{cases} P_{ij}(\boldsymbol{u})\,\Delta t + \cdots, & i \neq j, \\ 1 + \cdots, & i = j. \end{cases} \tag{11.19}$$

The return to be optimized will have two kinds of terms: (1) a component of the form $\int r_i(\boldsymbol{u})\,dt$ computed over times during which the state is fixed; (2) a component $\sum r'_{ij}(\boldsymbol{u})$ computed at a time at which the state changes.

We shall first assume that the horizon is fixed at $T$ and that the final state is free.

### *11.3.2 Solution of the Problem*

Let $\hat{R}_i(t)$ be the mathematical expectation of the optimal return when the initial state is $i$ at time $t$. For small $\Delta t$, the principle of optimality leads to

$$\hat{R}_i(t) = \operatorname*{opt}_{\boldsymbol{u} \in \mathscr{D}}\{r_i(\boldsymbol{u})\,\Delta t + \hat{R}_i(t + \Delta t) + \sum_{j \neq i} [r'_{ij}(\boldsymbol{u}) + \hat{R}_j(t + \Delta t)]P_{ij}(\boldsymbol{u})\,\Delta t\}. \tag{11.20}$$

Assuming $\hat{R}_i(t)$ to be continuous in $t$,

$$\hat{R}_i(t + \Delta t) = \hat{R}_i(t) + (\mathrm{d}\hat{R}_i(t)/\mathrm{d}t)\,\Delta t + \cdots. \tag{11.21}$$

Using this in (11.20) and passing to the limit results in

$$\mathrm{d}\hat{R}_i(t)/\mathrm{d}t + \operatorname*{opt}_{\boldsymbol{u} \in \mathscr{D}}\{r_i(\boldsymbol{u}) + \sum_{j \neq i} [r'_{ij}(\boldsymbol{u}) + \hat{R}_j(t)]P_{ij}(\boldsymbol{u})\} = 0. \tag{11.22}$$

Letting $\hat{\boldsymbol{R}}(t)$ be the vector with elements $\hat{R}_i(t)$, $\boldsymbol{C}(\boldsymbol{u})$ the vector with elements

$$C_i(\boldsymbol{u}) = r_i(\boldsymbol{u}) + \sum_{j \neq i} r'_{ij}(\boldsymbol{u})P_{ij}(\boldsymbol{u}),$$

and $\mathbf{P}(\boldsymbol{u})$ the matrix with elements $P_{ij}(\boldsymbol{u})$ for $i \neq j$, and 0 for $i = j$, then the relations (11.22) can be written

$$\mathrm{d}\hat{\boldsymbol{R}}(t)/\mathrm{d}t = -\operatorname*{opt}_{\boldsymbol{u} \in \mathcal{D}}\{\boldsymbol{C}(\boldsymbol{u}) + \mathbf{P}(\boldsymbol{u})\hat{\boldsymbol{R}}(t)\}. \tag{11.23}$$

Each row of this is to be optimized independently for $\boldsymbol{u} \in \mathcal{D}$. If the final state is free, the boundary condition is

$$\hat{\boldsymbol{R}}(T) = \boldsymbol{0}. \tag{11.24}$$

### *11.3.3 Practical Calculation of the Solution*

Since the differential system (11.23) has a terminal condition, rather than an initial condition, it is integrated in reverse time. Thus, letting

$$\tau = T - t,$$

there results

$$\hat{\boldsymbol{R}}(0) = \boldsymbol{0},$$

$$\mathrm{d}\hat{\boldsymbol{R}}(\tau)/\mathrm{d}\tau = \operatorname*{opt}_{\boldsymbol{u} \in \mathcal{D}}\{\boldsymbol{C}(\boldsymbol{u}) + \mathbf{P}(\boldsymbol{u})\hat{\boldsymbol{R}}(\tau)\}. \tag{11.25}$$

This system can be integrated numerically step by step, or using the analog computation scheme indicated in Fig. 11.1. For certain simple problems, the optimization units can be replaced by an equivalent nonlinearity, such as the signum function.

### *11.3.4 The Case of an Infinite Horizon, or an Unspecified Horizon*

With an infinite or unspecified horizon we suppose that the process and the criterion function are stationary. The optimal return is then independent of time.

A development analogous to that used in the earlier case leads to the following result, in which the term $\mathrm{d}\hat{R}/\mathrm{d}t$ no longer appears:

$$\operatorname*{opt}_{\boldsymbol{u} \in \mathcal{D}}\{\boldsymbol{C}(\boldsymbol{u}) + \mathbf{P}(\boldsymbol{u})\hat{\boldsymbol{R}}\} = 0. \tag{11.26}$$

Only those coordinates relating to states which are not in the terminal domain appear in the vector $\hat{\boldsymbol{R}}$.

The problems of existence and uniqueness of solutions, and of their practical calculation, can be treated the same as in the discrete case, and we shall not discuss them further.

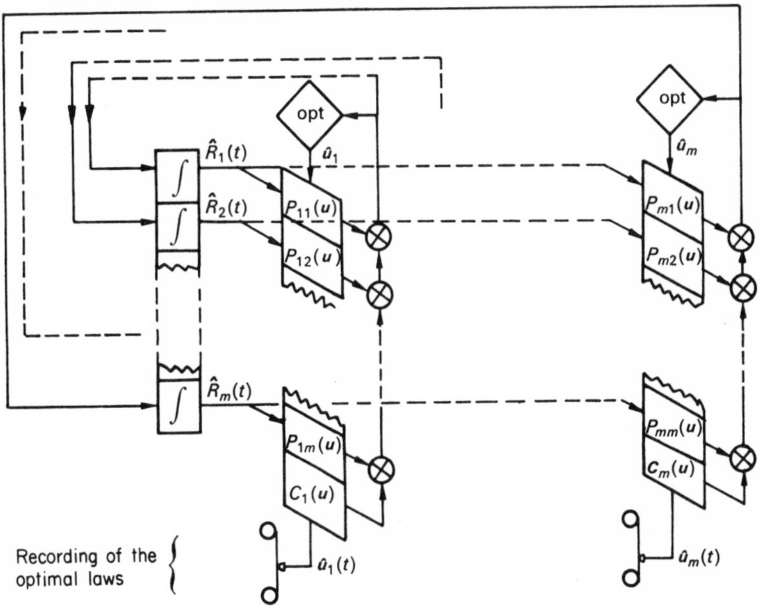


FIG. 11.1. Scheme for solution of Eq. (11.25).

# PART 5 Applications

In this final part of the book, we shall present a number of exercises in which the theoretical principles developed in the earlier chapters can be applied. It is necessary to be rather specific in these examples, particularly in the flow chart of the controller. However, due to their brevity, they are still rather far from examples of actual control problems. In addition, we have avoided examples which, after a short analysis, lead to a numerical solution, the result of which is some graphs or a table of numbers, neither of which would be of much interest so far as understanding the principles of the methods is concerned.

The first problem is elementary, and is intended to illustrate the general considerations developed in the first chapter.

In the second problem, which is also very simple, the establishment and solution of the optimal recurrence equation is studied. An analytical solution is possible, and the resulting optimal control law has a simple physical interpretation. In addition, it can be compared to the solution for the corresponding continuous problem.

The third example illustrates some of the numerical problems encountered in solving the optimal recurrence equation.

In the fourth example, the simplicity of the calculations needed to optimize a quadratic return in the case of a linear system is pointed out.

Control of a stochastic process is considered in the fifth example, and it is shown that depending on the nature of the criterion the solution is more or less complex.

In the sixth example, a minimum-time problem for a second-order nonlinear system is solved, by reducing it to a first-order discrete problem, by judicious formulation of the equations.

The seventh example deals with a continuous system. The partial differential equations are formulated starting with the corresponding discrete equations, and it is shown how a simple direct solution can be obtained by properly formulating the equations.

In the eighth exercise, the solution to a continuous problem is sought, using the partial differential equation of optimality. The solution found is compared with that obtained by using the maximum principle, and the use of the adjoint system in analog synthesis of the controller is investigated.

# Problem 1 | Introductory Example

In order to understand better the principle of optimality, which is the basis of the method of dynamic programming, we shall consider a simple problem. Although it is rather far from being an automatic control problem, it well illustrates the various steps in the solution of a problem using dynamic programming.

## P1.1 Problem A

Consider a smuggler who operates in the area around the intersection of the borders of three countries, denoted 1, 2, and 3. Each night he makes a single border crossing, and, for the moment, we assume he runs no risk of being surprised by the customs officials.

Passage from country $i$ to country $j$ brings an ultimate profit of $r_{ij}$ to the smuggler, as shown in Fig. P1.1. Starting in country $i = 1$, we wish to find the itinerary which leads to the maximum profit at the end of ten crossings.

| $i \backslash j$ | 1 | 2 | 3 |
|---|---|---|---|
| 1 | 0 | 10 | 7 |
| 2 | 1 | 0 | 4 |
| 3 | 8 | 2 | 0 |

FIG. P1.1. Table of elementary returns.

### *P1.1.1 Classical Solution*

Since at each passage the smuggler is faced with two alternatives, there are $2^{10} = 1024$ possible itineraries. A simple method, in theory at least, is to evaluate the profit for each possible itinerary and retain the largest. Even though the calculations required in this method would be long and tedious if carried out by hand, they are quite simple if an electronic computer is used. However, the amount of calculation increases rapidly as the number of passages $N$ increases, and, furthermore, the solution found for $N = 10$ is of no use in finding the solution for $N = 11$. In the same way, if the country of origin changes, the whole problem must be worked again.

### *P1.1.2 Solution by Dynamic Programming*

Three steps are involved in solving a problem by dynamic programming:

1. A more general class of problems is set up, which contains the problem of interest as a special case (invariant embedding).
2. The optimal return is considered as a general function of the variables which determine its value.
3. The principle of optimality is used to obtain a recurrence relation involving the optimal return function.

Let us illustrate these steps as they apply to our problem:

1. Rather than considering country 1 as the country of origin, and 10 passages, consider the general problem beginning in country $i$ and lasting $N$ passages.
2. The profit, or return, using an optimal itinerary is a function $R(i, N)$ which depends only on the country of departure $i$ and the number of passages $N$.
3. If the first decision $j$ is optimal, the smuggler finds himself in country $j$ with $N - 1$ passages remaining. For this new problem, he will use an optimal itinerary, and thereby gain a profit of precisely $R(j, N - 1)$. If the function $R(j, N - 1)$ is continuous, the best initial decision is that value of $j$ which maximizes the total return

$$r_{ij} + R(j, N - 1).$$

The functions $R(j, N)$ are thus related by the recursion

$$R(i, N) = \max_{j \in \{1, 2, 3\}} [r_{ij} + R(j, N - 1)]. \tag{P1.1}$$

The above reasoning is in accord with the principle of optimality: Whatever the initial state, if the first decision belongs to an optimal policy, then the policy relative to the problem starting from the state which resulted from that decision should be optimal.

### *P1.1.3 Numerical Solution of the Recurrence Relation*

Using Eq. (P1.1), the optimal return can be computed step by step for increasing $N$, beginning with the obvious solution

$$R(i, 0) = 0,$$

for $N = 0$. At each step, for each $i$ the corresponding optimal $j$ is recorded, which allows subsequent construction of the entire optimal policy.

For the problem of the smuggler, we shall represent the various optimal returns for some $N$ by

$$\begin{array}{c} R(1, N) \\ R(2, N) \\ R(3, N). \end{array}$$

These tables for various $N$ will be connected with arrows, representing the various optimal decisions. A connected chain of arrows then forms an optimal policy for some particular initial state.

Solution of Eq. (P1.1) in this case leads to the scheme of Fig. P1.2. No

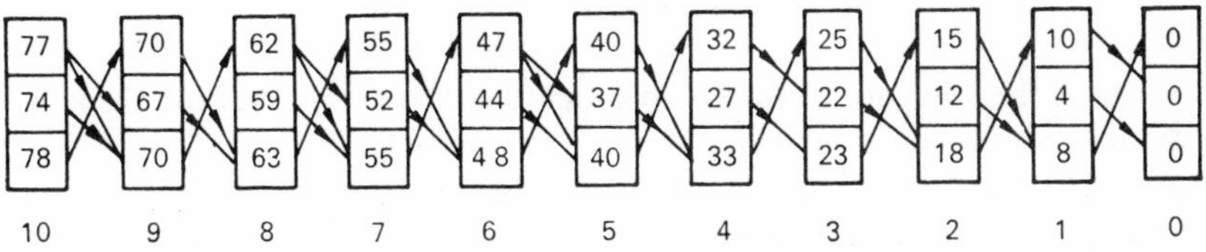


FIG. P1.2. Solution of the optimality equation.

difficulties arise here, and the solution can be easily calculated by hand, showing the advantage of the dynamic programming method. Further, the figure indicates simultaneously the solutions for all $i$ and for all $N \leq 10$.

### *P1.1.4 Remarks*

**Remark 1.** For $N = 6, 8, 10, \ldots$ it is possible to receive the optimal return using more than one decision. This illustrates the fact that the optimal policy is not necessarily unique. Thus, for the initial problem ($i = 1, N = 10$), each of the following itineraries is optimal:

1–2–3–1–3–1–3–1–3–1–2,
1–3–1–2–3–1–3–1–3–1–2,
1–3–1–3–1–2–3–1–3–1–2,
1–3–1–3–1–3–1–2–3–1–2.

It is easy to check that these itineraries are all equivalent, because each terminates in state 2, and each consists of two passages from 1 to 2, for a return of $2 \times 10 = 20$; three passages from 1 to 3 for a return of $3 \times 7 = 21$; one passage from 2 to 3 for a return of $1 \times 4 = 4$; and four passages from 3 to 1 for a return of $4 \times 8 = 32$. In each case, the total return is $R(1, 10) = 77$.

**Remark 2.** For $N \geq 5$, a periodicity is evident in the optimal control sequence, of the form shown in Fig. P1.3. This constitutes a steady state,

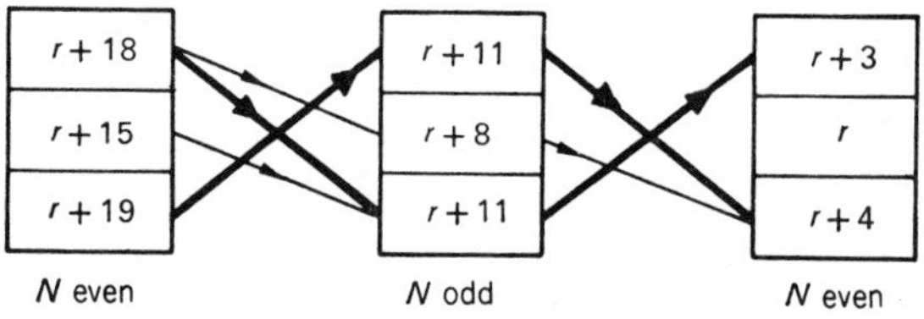


FIG. P1.3. Periodic form of the solution.

in which the influence of the last passage is felt, for large $N$, only through the parity of $N$, even or odd.

The trajectory of the steady state, indicated by heavy arrows in Fig. P1.4, which corresponds to Fig. P1.3, leads to an average return of $15/2 = 7.5$.

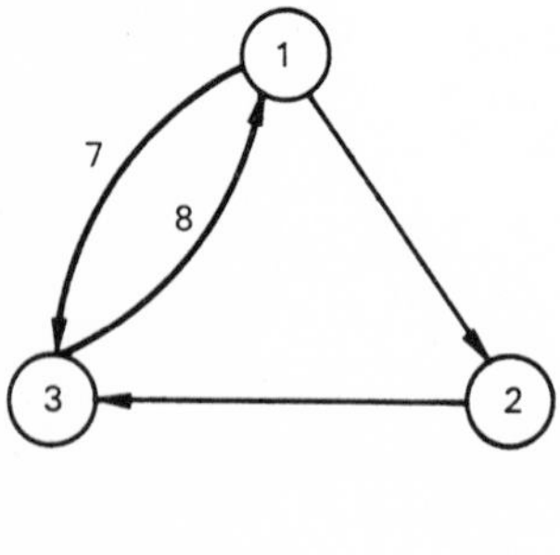


FIG. P1.4. The smuggler's route in the periodic mode.

**Remark 3.** A greedy smuggler will go each time into whatever country appears most profitable immediately. Examination of the profit matrix of the $r_{ij}$ shows that this leads to the itinerary shown in Fig. Pl.5, which

FIG. P1.5. The route of the greedy smuggler.

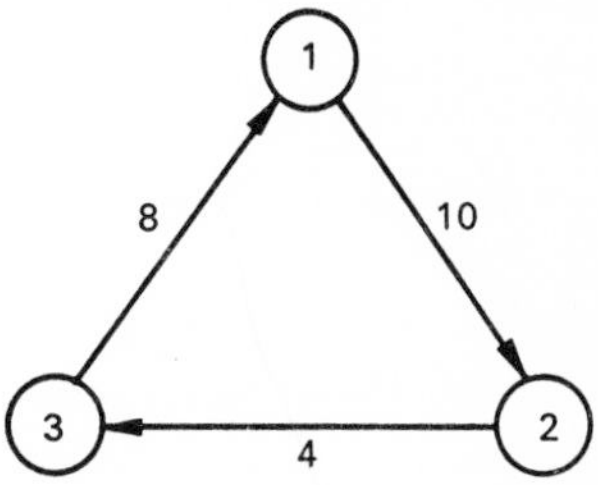


corresponds to a return for each three passages of $10 + 8 + 4 = 22$, which is to say an average return of $22/3 = 7.33$. This is less than the return realized by an optimal smuggler.

## P1.2 Problem B

Let us consider the same problem as in Section P1.1, except that now the smuggler wishes to be in country 1 as a result of the last passage of his itinerary.

### *P1.2.1 Solution by Dynamic Programming*

The recurrence relation (P1.1) remains valid. The final decision, however, is now forced in that it must lead into country 1. Thus Eq. (P1.1) can now be used only for $N > 1$. The resulting solution is shown in Fig. P1.6.

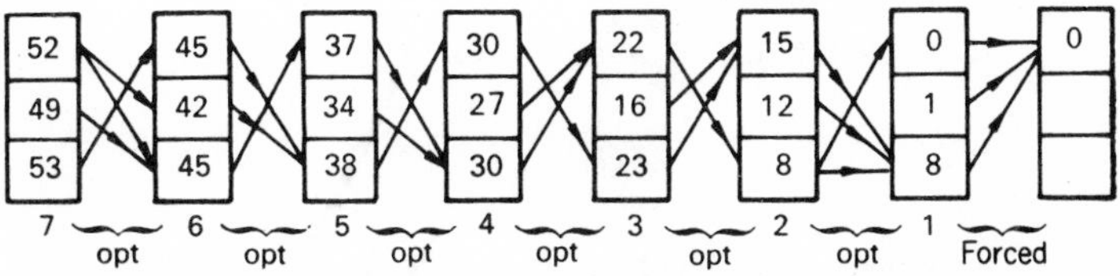


FIG. P1.6. Optimal solution for the smuggler with a terminal constraint.

### *P1.2.2 Remarks*

**Remark 1.** This constraint on the final state results in a decrease in the optimal return. For example, for $i = 1$ and $N = 7$, the smuggler now receives a return of 52, rather than 55 as would be the case if the final state were free.

**Remark 2.** The same steady state is established for $N > 5$.

## P1.3 Problem C

Now the smuggler risks capture during a passage, and seizure of the goods he carries during that passage. However, the probability that he will be apprehended depends on which frontier he crosses, and on which direction he travels. These probabilities are indicated in the matrix of Fig. P1.7, with $p_{ij}$ being the probability that he will pass from $i$ to $j$ safely.

| | 1 | 2 | 3 |
|---|---|---|---|
| 1 | 1 | 0.6 | 0.6 |
| 2 | 0.9 | 1 | 0.8 |
| 3 | 0.6 | 0.9 | 1 |

FIG. P1.7. Probabilities of free passage for the case of guarded frontiers.

### *P1.3.1 Method of Solution*

The stopping time of the process is now unknown. For a given policy and initial state, the return which will be realized up till the time of capture is a random variable. Thus the optimal return can be defined only in terms of the mathematical expectation of the return. This is a function which depends only on the state, $R(i)$.

Using the principle of optimality, the optimal return $R(i)$ can be described by

$$R(i) = \max_{j \neq i}[p_{ij}\{r_{ij} + R(j)\}]. \qquad \text{(P1.2)}$$

Here $p_{ij}r_{ij}$ is the expected return for passage from $i$ to $j$, and $p_{ij}R(j)$ is the expected future gain until capture. Rather than a recurrence relation, we have here obtained an implicit equation. This follows since after each decision is carried out (if the smuggler was not caught), his expected future gain is unchanged from the situation before he made his decision.

Letting $R(1) = x$, $R(2) = y$, and $R(3) = z$, Eq. (P1.2) becomes

$$\begin{aligned} x &= \max[6 + 0.6\,y \,|\, 4.2 + 0.6\,z], \\ y &= \max[0.9 + 0.9\,x \,|\, 3.2 + 0.8\,z], \\ z &= \max[4.8 + 0.6\,x \,|\, 1.8 + 0.9\,y]. \end{aligned} \tag{P1.3}$$

*P1.3.2 Numerical Solution*

To solve the set of implicit equations (P1.3), an iterative method can be used. It was shown in Chapter 6 that such a method converges. System (P1.3) then leads to

$$\begin{aligned} x_{n+1} &= \max[6 + 0.6\,y_n \,|\, 4.2 + 0.6\,z_n], \\ y_{n+1} &= \max[0.9 + 0.9\,x_n \,|\, 3.2 + 0.8\,z_n], \\ z_{n+1} &= \max[4.8 + 0.6\,x_n \,|\, 1.8 + 0.9\,y_n]. \end{aligned}$$

Initializing with $x_0 = y_0 = z_0 = 1$, this results in

$$\begin{bmatrix} 1 \\ 1 \\ 1 \end{bmatrix} \to \begin{bmatrix} \mathbf{6.6} & 4.8 \\ 1.8 & \mathbf{4} \\ \mathbf{5.4} & 2.7 \end{bmatrix} \to \begin{bmatrix} 6.6 \\ 4 \\ 5.4 \end{bmatrix} \to \begin{bmatrix} \mathbf{8.4} & 7.74 \\ 6.84 & \mathbf{7.52} \\ \mathbf{8.04} & 6.4 \end{bmatrix} \to \begin{bmatrix} 8.4 \\ 7.52 \\ 8.04 \end{bmatrix}$$

$$\to \begin{bmatrix} \mathbf{10.51} & 9.02 \\ 8.46 & \mathbf{10.63} \\ \mathbf{9.84} & 8.56 \end{bmatrix}.$$

Convergence is thus immediate, and the optimal policy is $1 \to 2$, $2 \to 3$, $3 \to 1$. The corresponding returns can be calculated by using this policy in (P1.3) to obtain

$$\begin{aligned} x &= 6 + 0.6\,y & & & x &= 14.36, \\ y &= 3.2 + 0.8\,z & &\to & y &= 13.93, \\ z &= 4.8 + 0.6\,x & & & z &= 13.42. \end{aligned}$$

This result satisfies (P1.3), and is indeed optimal.

**Remark 1.** The policy found here is in fact identical to that used by the greedy smuggler.

**Remark 2.** The expected number of passages up till capture, using the optimal policy, can be calculated easily. Letting $N(1)$, $N(2)$, and $N(3)$ be the expected stopping times for the three possible initial states, we have

$$\begin{aligned} N(1) &= 0.6[1 + N(2)] & & N(1) = 1.92, \\ N(2) &= 0.8[1 + N(3)] & \rightarrow \quad & N(2) = 2.2, \\ N(3) &= 0.6[1 + N(1)] & & N(3) = 1.75. \end{aligned}$$

These values are rather small because the customs agents are quite active. This also explains why the "greedy" policy is optimal.

## P1.4 Conclusion

For this elementary problem, we have seen that the method of dynamic programming yields a solution much more rapidly than does the classical combinatorial method. The basic ideas of dynamic programming are very simple. We have also brought out the concept of a steady state behavior as the horizon tends towards infinity. We have shown that introducing terminal conditions on the state causes no difficulties.

Finally, we have shown that the identical method can be applied whether the problem is deterministic or stochastic.

# Problem 2 Minimum Use of Control Effort in a First-Order System

## P2.1 Statement of the Problem

Consider the system described by the first-order equation

$$x_{n+1} = ax_n + u_n, \tag{P2.1}$$

where $x_n$ is the state at time $n$ and $u_n$ is the control applied during sampling period number $n$. It is desired to transfer the system from some initial state $X$ to the final state $X_f = 0$ in $N$ steps, while minimizing the criterion $\sum_0^{N-1} |u_n|$. If the instantaneous fuel consumption is proportional to $|u|$, for example, this criterion is the total fuel used during control of the process, and we have a minimum-fuel problem.

We shall assume first that $u_n$ is unconstrained, and discuss the form of the optimal policy for various values of $a$. The constraint $|u| \leq 1$ will then be applied, and the domains which are controllable to the origin found. The optimal policy, for $|a| < 1$, is then found, and the flow chart of the numerical optimal controller presented. Finally, we shall consider the case in which the discrete model above corresponds to numerical control of the continuous process

$$\dot{y} = -\alpha y + u. \tag{P2.2}$$

We shall establish formulas which allow passage from one model to the other, and study the way in which the optimal control law changes as the sampling step $\Delta$ tends to zero. The results will be justified using the results of the classical calculus of variations.

## P2.2 The Unconstrained Case

### *P2.2.1 Formulation of the Equations*

Since the process is stationary, the optimal control law depends only on the state $x_n$, and the number of sampling periods, $m = N - n$, remaining until the stopping time $N$. We shall thus count time backward from the terminal time, and write the state equation as

$$x_{m-1} = ax_m + u_m . \tag{P2.3}$$

The terminal condition is $x_0 = 0$ in this notation.

Letting $R_m(X)$ be the optimal return relative to the initial state $X$, and with $m$ sampling periods remaining, application of the principle of optimality leads to

$$R_m(X) = \min_u \{|u| + R_{m-1}(aX + u)\} . \tag{P2.4}$$

### *P2.2.2 Initialization of the Optimal Recurrence Equation*

Since the final state, $x_0$, is forced to be the origin, for $m = 1$ relation (P2.3) becomes

$$0 = ax_1 + u_1 \rightarrow u_1 = -ax_1 . \tag{P2.5}$$

Since $u_1$ is thus fixed, the minimization operator in the recurrence relation (P2.4) is ineffective, and we have for $m = 1$ that

$$R_1(X) = |u_1| = |a|\ |X| . \tag{P2.6}$$

Thus (P2.4) will be used for $m > 1$ with (P2.6) as initial condition.

### *P2.2.3 Solution of the Recurrence Relation*

In solving the recurrence relation (P2.4), we shall begin with the case $m = 2$, and then show that the result obtained can be extended to arbitrary $m$ without difficulty.

For $m = 2$, the relation (P2.4) becomes

$$R_2(X) = \min_u [|u| + |a|\ |aX + u|] = \min_u f(X, u) . \tag{P2.7}$$

In order to carry out the minimization, we distinguish two cases of $f(X, u)$. First, for $|a| > 1, f(X, u)$ is as shown in Fig. P2.1, since

$$|a| > 1 \Rightarrow |a|\ |X| < |a|^2 |X|.$$

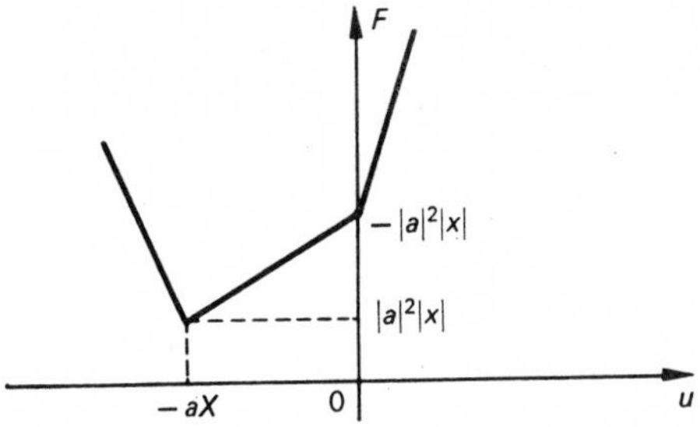


FIG. P2.1. The function $f$ of equation (P2.7).

In this case,

$$\hat{u}_2 = -ax_2,$$

$$R_2(X) = |a|\ |X| = R_1(X) \quad \Rightarrow \quad \hat{u}_1 = 0.$$

It follows that

$$R_m(X) = |a|\ |X|, \qquad \forall m, \tag{P2.8}$$

so that the optimal policy, starting from $X_N$, is simply

$$\hat{u}_N = -aX_N, \quad \hat{u}_{N-1} = 0, \quad \ldots, \quad \hat{u}_1 = 0.$$

In the case that $|a| < 1$, Fig. P2.1 is modified, to take account of the fact that

$$|a| < 1 \Rightarrow |a|\ |X| > |a|^2 |X|.$$

In this case,

$$\hat{u}_2 = 0 \Rightarrow R_2(X) = |a|^2 |X|,$$

so that

$$R_m(X) = |a|^m |X|. \tag{P2.9}$$

The optimal policy is now

$$\hat{u}_N = 0, \quad \hat{u}_{N-1} = 0, \quad \ldots, \quad \hat{u}_1 = -|a|^N X_N = -aX_1.$$

### *P2.2.4 Interpretation of the Optimal Policies*

The following physical reasoning can be used to explain the above two results. For $|a| < 1$, the system is stable, and can be allowed to evolve unhindered towards the origin, until the last instant, at which time the system is forced to the origin with the last control. For $|a| > 1$, the system is unstable, and it is forced to the origin at once, using the first control signal.

## P2.3 The Constrained Case

### *P2.3.1 Calculation of the Controllable Regions*

A state $X$ is controllable (to the origin) in $m$ steps, in this case, if there exist $u_m, \ldots, u_1$, each of absolute value less than 1, such that

$$x_0 = 0 = \sum_{i=1}^{m} a^{i-1} u_i + a^m X.$$

Thus

$$X = -\sum_{1}^{m} a^{i-1-m} u_i. \tag{P2.10}$$

The region $\mathcal{X}_m$ of controllable states is then given by

$$\mathcal{X}_m = \left\{ X = -\sum_{i=1}^{m} a^{i-1-m} u_i \,\middle|\, |u_i| \leq 1 \right\}. \tag{P2.11}$$

It is easy to see that the domain (P2.11) is a line segment $[-\varepsilon_m, \varepsilon_m]$, the limits of which are attained for $u_i$ such that each term of the sum is maximum (or minimum). Thus

$$\max\{-\sum a^{i-1-m} u_i\} = \sum [\max_{u_i}(-a^{i-1-m} u_i)],$$

$$\min\{-\sum a^{i-1-m} u_i\} = \sum [\min_{u_i}(-a^{i-1-m} u_i)],$$

with each maximum (or minimum) being computed independently, so that

$$\hat{u}_i = \pm 1.$$

Thus

$$\varepsilon_m = \sum_{1}^{m} |a|^{i-1-m} = |a|^{-1} + |a|^{-2} + \cdots + |a|^{-m},$$

$$\varepsilon_m = |a|^{-1}(|a|^{-m} - 1)/(|a|^{-1} - 1) = (|a|^{-m} - 1)/(1 - |a|). \quad \text{(P2.12)}$$

A recurrence relation for these bounds can be obtained by writing

$$\varepsilon_m = |a|^{-1}[1 + (|a|^{-1} + |a|^{-2} + \cdots + |a|^{-(m-1)})],$$

$$\varepsilon_m = |a|^{-1}[1 + \varepsilon_{m-1}], \qquad \text{with} \quad \varepsilon_0 = 0,$$

$$\varepsilon_{m-1} = |a|\varepsilon_m - 1. \quad \text{(P2.13)}$$

The last of these expressions indicates that the $\varepsilon_m$ can be calculated successively in time using the system shown in Fig. P2.2, initialized for $m = N$ with

$$\varepsilon_{\text{initial}} = (|a|^{-N} - 1)/(1 - |a|).$$

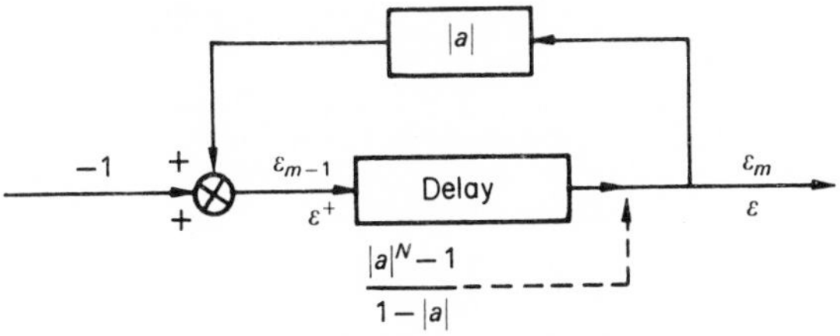


FIG. P2.2. Iterative computation of the controllable domain.

For $m \to \infty$, the boundaries of the controllable regions are

$$\varepsilon_\infty = \infty, \qquad |a| < 1,$$

$$\varepsilon_\infty = 1/(|a| - 1), \qquad |a| > 1.$$

### *P2.3.2 The Optimal Recurrence Equation*

In this case, the recurrence relation for the optimal returns differs from (P2.4) in that $u$ is bounded, so that

$$R_m(X) = \min_{-1 \le u \le 1} \{|u| + R_{m-1}(aX + u)\}. \quad \text{(P2.14)}$$

This relation is initialized by using the terminal condition. Thus, for

$m = 1$, $u = -aX_1$, and $R_1(X) = |a|\ |X|$. However, since $|u| \leq 1$, $R_1(X)$ is only defined for

$$-1/a < X_1 < 1/a = \varepsilon_1,$$

which is to say on the corresponding controllable domain.

### *P2.3.3 Solution of the Recurrence Relation for $m = 2$*

In the case of $m = 2$, the relation (P2.14) becomes

$$R_2(X) = \min_{|u| \leq 1} [|u| + |a|\ |aX + u|], \tag{P2.15}$$

with

$$|aX + u| < 1/a = \varepsilon_1.$$

The minimization in (P2.14) can be carried out using the graphical construction shown in Fig. P2.3. Assuming that $|a| < 1$, the following calculations result:

$$\left.\begin{array}{l} -aX + \varepsilon_1 > 0 \\ -aX - \varepsilon_1 < 0 \end{array}\right\} \Rightarrow \hat{u} = 0,$$

$$-1 < -aX + \varepsilon_1 < 0 \Rightarrow \hat{u} = -aX + \varepsilon_1,$$

$$0 < -aX - \varepsilon_1 < 1 \Rightarrow \hat{u} = -aX - \varepsilon_1.$$

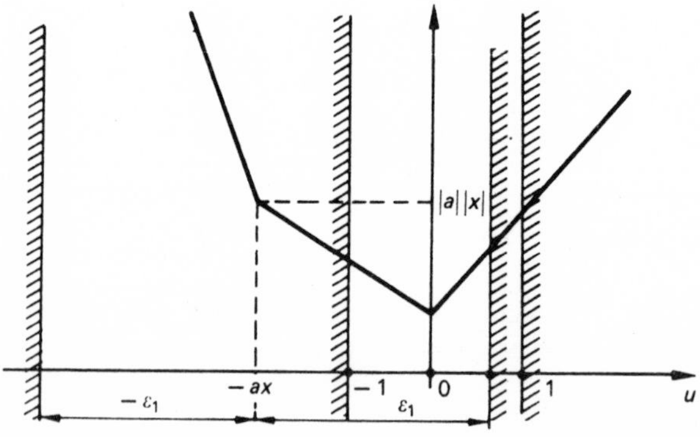


FIG. P2.3. The function to be minimized in Eq. (P2.15).

These can be rearranged as

$$-\varepsilon_2 < X < -\varepsilon_1/|a|, \qquad \hat{u} = -aX + \varepsilon_1,$$
$$-\varepsilon_1/|a| < X < \varepsilon_1/|a|, \qquad \hat{u} = 0,$$
$$\varepsilon_1/|a| < X < \varepsilon_2, \qquad \hat{u} = -aX - \varepsilon_1.$$

The corresponding return $R_2(X)$, which is defined only on $[-\varepsilon_2, \varepsilon_2]$, is shown in Fig. P2.4.

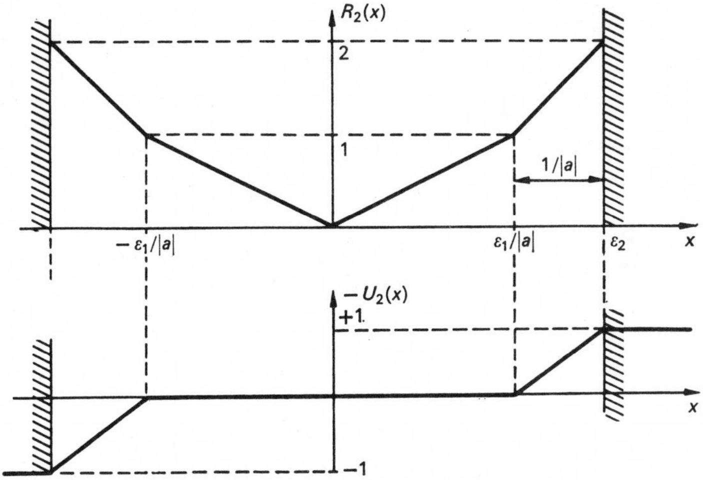


FIG. P2.4. The optimal return function and optimal control law for two steps of the process.

### *P2.3.4 General Solution*

In order to extend the above result to arbitrary $m$ we shall show that $R_m(X)$ is a convex function made up of straight-line segments, defined on $[-\varepsilon_m, \varepsilon_m]$, and at the limits of this interval has a slope which is just $a$, with $|a| < 1$. Since these properties hold for $m = 2$, as found above, we shall proceed inductively, and assume that they are true for $m - 1$. We need then to show that they hold true for $m$.

The search for $R_m(X)$ leads to the construction shown in Fig. P2.5. Since $R_{m-1}$ is a convex function, its slope is always less than the slope at the boundaries of the region of definition and hence less than one. Thus the arc $AB$ is monotonically decreasing, as shown in the figure. In the same

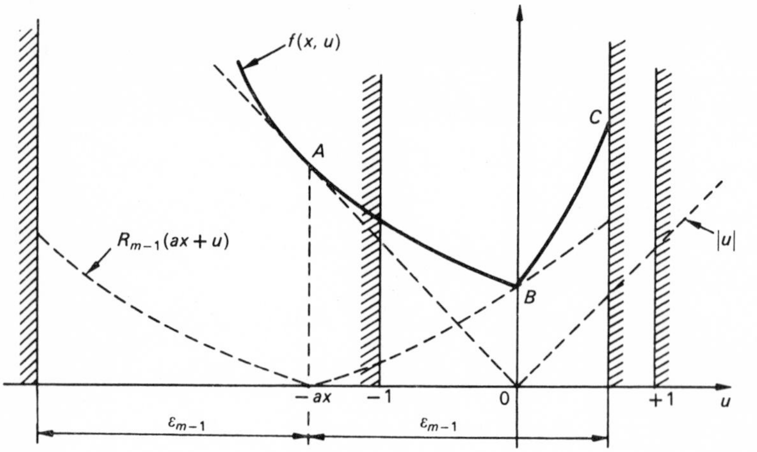


FIG. P2.5. Optimal return function in general.

way, $BC$ is monotonically increasing. Thus, depending on the value of $X$, we have

$$\left.\begin{array}{l} -aX+\varepsilon_{m-1}>0 \\ -aX-\varepsilon_{m-1}<0 \end{array}\right. \Rightarrow \hat{u}_m = 0,$$

$$-1 < -aX+\varepsilon_{m-1} < 0 \Rightarrow \hat{u}_m = -aX+\varepsilon_{m-1},$$

$$0 < -aX-\varepsilon_{m-1} < 1 \Rightarrow \hat{u}_m = -aX-\varepsilon_{m-1}.$$

The optimal policy is thus

$$-\varepsilon_m < X < -\varepsilon_{m-1}/|a| \Rightarrow \hat{u}_m = -aX+\varepsilon_{m-1},$$

$$-\varepsilon_{m-1}/|a| < X < \varepsilon_{m-1}/|a| \Rightarrow \hat{u}_m = 0,$$

$$\varepsilon_{m-1}/|a| < X < \varepsilon_m \Rightarrow \hat{u}_m = -aX-\varepsilon_{m-1}.$$

The corresponding optimal return is then given by

$$-\varepsilon_{m-1}/|a| < X < \varepsilon_{m-1}/|a| \Rightarrow R_m(X) = R_{m-1}(aX),$$

$$\varepsilon_{m-1}/|a| < X < \varepsilon_m \Rightarrow R_m(X) = |aX+\varepsilon_{m-1}| + R_{m-1}(\varepsilon_{m-1}),$$

$$R_m(X_{\max}) = R_m(\varepsilon_m) = 1 + R_{m-1}(\varepsilon_{m-1}) \Rightarrow R_m(\varepsilon_m) = m.$$

Thus $R_m(X)$ is composed of three convex arcs.

At the junction of these segments ($aX = \varepsilon_{m-1}$), the first has slope

$$\partial R_m(X)/\partial x = \partial R_{m-1}(aX)/\partial x = a\,[\partial R_{m-1}(y)/\partial y]_{y=\varepsilon_{m-1}},$$

which is of absolute value $|a|^2$, and the second has slope

$$\partial R_m(X)/\partial X = \partial|aX + \varepsilon_{m-1}|/\partial x,$$

of absolute value $|a| > |a|^2$, since $|a| < 1$. This establishes the convexity of $R_m(X)$.

It is easily verified that $R_m(X)$ is only defined on $[-\varepsilon_m, \varepsilon_m]$, and that its slope at the ends of this interval is

$$\partial|aX + \varepsilon_{m-1}|/\partial x$$

which is of absolute value $|a|$. In fact, $R_m(X)$ is made up of line segments corresponding to $\Delta R = 1$ for $\Delta X$ which are successively

$$1/|a|^m, \quad 1/|a|^{m-1}, \quad \ldots, \quad 1/|a|.$$

The optimal control policy determined by formula (P2.10) is shown in Fig. P2.6.

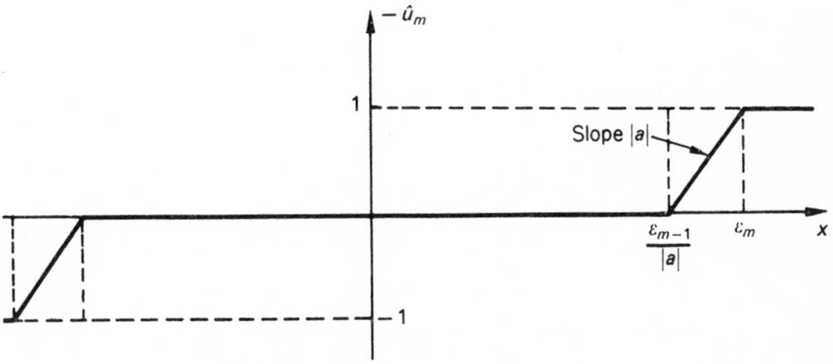


FIG. P2.6. Optimal control law in general.

### *P2.3.5 Synthesis of the Numerical Controller*

The above optimal control law leads to the flow chart shown in Fig. P2.7.

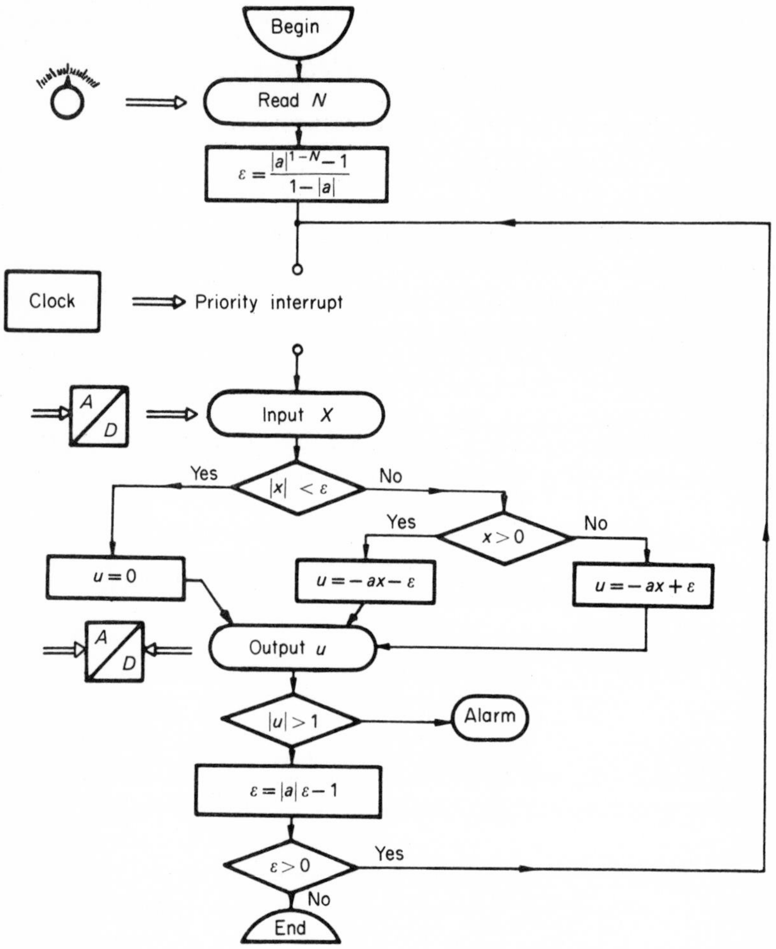


FIG. P2.7. Optimal controller of the first-order system.

## P2.4 Passage to a Continuous System

Let us consider the continuous linear system

$$\dot{y} = -\alpha y + n, \qquad |u| < 1 . \tag{P2.16}$$

If this system is controlled using a control interval $\Delta$, it can be represented by the discrete model

$$y^{+} = \mathrm{e}^{-\alpha\Delta} y + (1 - \mathrm{e}^{-\alpha\Delta}) u/\alpha . \tag{P2.17}$$

Then defining $x$ by

$$y = (1 - e^{-\alpha\Delta})x/\alpha, \tag{P2.18}$$

there results the discrete system considered above, with

$$a = e^{-\alpha\Delta}. \tag{P2.19}$$

In terms of $y$, the control law of Fig. P2.6 becomes that shown in Fig. P2.8.

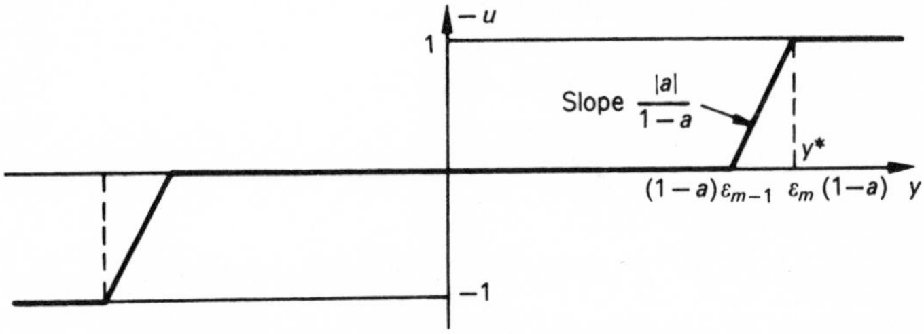


FIG. P2.8. Continuous optimal control law.

Let us now consider the limit as $m \to \infty$, holding $m\Delta = T =$ constant. The slope of the control law then behaves as

$$e^{-\alpha\Delta}/(1 - e^{-\alpha\Delta}) \to \infty,$$

and the control becomes discontinuous at the point

$$y^* = \lim(1 - a)\varepsilon_m = e^{-\alpha m\Delta} - 1 = e^{-\alpha T} - 1. \tag{P2.20}$$

It can be seen that this control indeed leads to satisfaction of the terminal conditions (see Fig. P2.9).

In addition, using the maximum principle it can be shown that this control, of the type (0, −1) or (0, +1), extremizes the criterion $\int |u|\, dt$. The canonical system is

$$\begin{aligned} \dot{y} &= -\alpha y + u, \\ \dot{\psi} &= \alpha\psi \to \psi(t) = e^{\alpha t}\psi_0, \end{aligned} \tag{P2.21}$$

with the Hamiltonian

$$\mathcal{H} = -|u| + \psi(-\alpha y + u). \tag{P2.22}$$

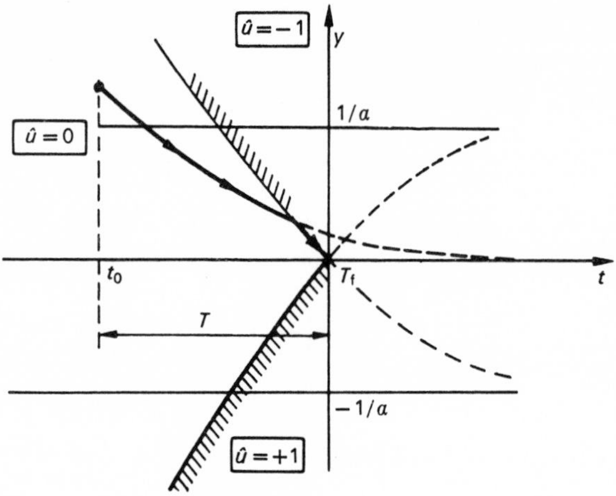


FIG. P2.9. State space for the continuous first-order system.

This Hamiltonian is maximized for

$$\hat{u} = 0, \qquad -1 < \psi < 1,$$

$$\hat{u} = -1, \qquad \psi < -1,$$

$$\hat{u} = +1, \qquad \psi > +1.$$

Since $\psi(t)$ is monotonic, and $\psi(\infty) = 0$, the optimal control can only be of the form $(0, -1)$ or $(0, +1)$, which is precisely that illustrated in Fig. P2.9.

# Problem 3 | Optimal Tabulation of Functions

## P3.1 Statement of the Problem

Suppose that the operation $y = f(x)$ must be programmed into a controller, where $f$ is a complicated analytic function, and with $x$ lying on some interval $[A, B]$. In order that this computation may be made rapidly, and with a reduced amount of memory required, suppose that it has been decided to approximate $f(x)$ by polynomials of degree $k$ over contiguous intervals $[x_n, x_{n+1}]$. If there are $N$ such segments, it is necessary to store $N - 1$ values $x_n$ and $N(k + 1)$ polynomial coefficients, for a total storage requirement of $(k + 2)N - 1$ words of memory.

Suppose further that $kN$ is fixed by the system requirements, and that the points $x_n$ and polynomial coefficients $\alpha_m{}^i$ are to be found such that the approximation to $f(x)$ is best, in the sense of some criterion

$$\int_A^B \mathscr{C}[f(x) - f^*(x)]\,\mathrm{d}x,$$

where $f^*$ is the approximate value computed for $y$.

After presenting the flow chart of a program for computing $f^*(x)$, we shall establish the optimal recurrence equation for the solution of this problem. We shall then examine the particular case of a quadratic criterion, and outline the optimization process for the polynomial coefficients, using Legendre polynomials.

## P3.2 Flow Chart for Evaluation of the Function

The flow chart of the calculation of $f^*(x)$ is shown in Fig. P3.1. There are required $2(k + 1)$ multiplications, $k + 1$ additions, and $N + k + 1$ comparisons.

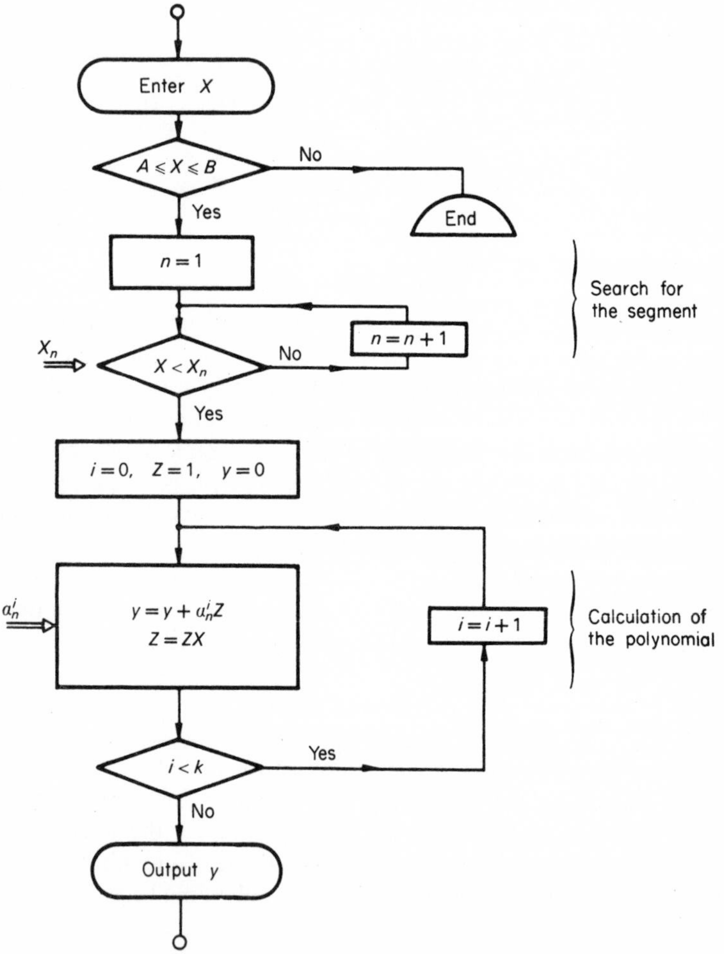


FIG. P3.1. Function evaluation based on Legendre polynomials.

## P3.3 Equation Formulation

In order to solve the problem relative to approximation of $f(x)$ by polynomials of degree $k$ with the interval $[A, B]$ segmented into $N$ segments, we shall consider the family of problems relative to approximation of $f(x)$ by polynomials of fixed degree $k$ with the interval $[A, y]$, $A \leq y < B$, segmented into an arbitrary number $m$ of segments.

For a given $f(x)$, the minimum of the criterion

$$\int_A^y (f(x) - f^*(x))\,\mathrm{d}x$$

only depends on $y$ and $m$, and is thus some function $\varphi_m(y)$. The functions $\varphi_m(y)$ and $\varphi_{m-1}(y)$ can be related by segmenting the interval $[A, y]$ into an interval $[A, z]$, with $A \leq z < y$, over which an approximation with $m - 1$ segments is used, and an interval $[z, y]$, over which the polynomial $\sum \alpha_m{}^i x^i$ is used. By the nature of the criterion, the principle of optimality yields

$$\varphi_m(y) = \min_{A \leq z < y \text{ and } \alpha_m{}^i} \left\{ \varphi_{m-1}(z) + \int_z^y \mathscr{C}[f(x) - \sum_i \alpha_m{}^i x^i]\,\mathrm{d}x \right\}, \tag{P3.1}$$

or

$$\varphi_m(y) = \min_{A \leq z < y} \left\{ \varphi_{m-1}(z) + \min_{\alpha_m{}^i} \left[ \int_z^y \mathscr{C}[f(x) - \sum_i \alpha_m{}^i x^i]\,\mathrm{d}x \right] \right\}. \tag{P3.2}$$

Minimization with respect to the $\alpha_m{}^i$ can be carried out separately to yield

$$\Delta(z, y) = \min_{\alpha_m{}^i} \left[ \int_z^y \mathscr{C}(f(x) - \sum_i \alpha_m{}^i x^i)\,\mathrm{d}x \right]. \tag{P3.3}$$

Then (P3.2) becomes

$$\varphi_m(y) = \min_{A \leq z < y} \{\varphi_{m-1}(z) + \Delta(z, y)\}. \tag{P3.4}$$

Solution of (P3.3) yields $\hat{\alpha}_m{}^i = \gamma_m{}^i(y, z)$, and solution of (P3.4) then results in $z = g_m(y)$. Using these, the various $x_n$ and $\hat{\alpha}_n{}^i$ can be calculated for the given $N$ and $B$:

$$\begin{aligned} x_{N-1} &= g_N(B) \to \hat{\alpha}_N{}^i = \gamma_N{}^i(B, x_{N-1}), \\ &\vdots \\ x_{m-1} &= g_m(x_m) \to \alpha_m{}^i = \gamma_m{}^i(B, x_{m-1}). \end{aligned} \tag{P3.5}$$

## P3.4 Case of a Quadratic Criterion

### *P3.4.1 Direct Formulation of the Equations*

Let us suppose now that the criterion is of the form

$$\mathscr{C} = \left[f(x) - \sum_i \alpha_m{}^i x^i\right]^2, \tag{P3.6}$$

$$R = \int_z^y \left[f(x) - \sum_i \alpha_m{}^i x^i\right]^2 \, dx. \tag{P3.7}$$

In order that $R$ be minimum, it is necessary that

$$\partial R/\partial \alpha_j = 2\int_z^y x^j \left[f(x) - \sum_i \alpha_m{}^i x^i\right] dx = 0. \tag{P3.8}$$

Defining

$$F_j(t) = \int_A^t f(x)x^j \, dx \tag{P3.9}$$

and

$$\int_A^t x^k \, dx = (t^{k+1} - A^{k+1})/(k+1) = \sigma_k(t), \tag{P3.10}$$

this can be written as the linear system

$$\begin{bmatrix} F_0(y) - F_0(z) \\ \vdots \\ F_i(y) - F_i(z) \\ \vdots \\ F_k(y) - F_k(z) \end{bmatrix}$$

$$= \begin{bmatrix} \sigma_0(y) - \sigma_0(z), \sigma_1(y) - \sigma_1(z), & \ldots, \sigma_k(y) - \sigma_k(z) \\ \vdots & \\ \sigma_i(y) - \sigma_i(z), \sigma_{i+1}(y) - \sigma_{i+1}(z), & \ldots, \sigma_{i+k}(y) - \sigma_{i+k}(z) \\ \vdots & \\ \sigma_k(y) - \sigma_k(z), \sigma_{k+1}(y) - \sigma_{k+1}(z), & \ldots, \sigma_{2k}(y) - \sigma_{2k}(z) \end{bmatrix} \begin{bmatrix} \hat{\alpha}_m{}^0 \\ \vdots \\ \hat{\alpha}_m{}^i \\ \vdots \\ \hat{\alpha}_m{}^k \end{bmatrix}. \tag{P3.11}$$

In matrix form, this is

$$\mathbf{F}(y) - \mathbf{F}(z) = [\boldsymbol{\sigma}(y) - \boldsymbol{\sigma}(z)]\hat{\boldsymbol{\alpha}}_m, \qquad \text{(P3.12)}$$

with solution

$$\hat{\boldsymbol{\alpha}}_m = [\boldsymbol{\sigma}(y) - \boldsymbol{\sigma}(z)]^{-1}[\mathbf{F}(y) - \mathbf{F}(z)]. \qquad \text{(P3.13)}$$

Since

$$\Delta(z, y) = \int_z^y f^2(x)\,\mathrm{d}x - 2\sum \alpha_m{}^i \int_z^y x^i f(x)\,\mathrm{d}x + \sum_i \sum_j \alpha_m{}^i \alpha_m{}^j \int_z^y x^{i+j}\,\mathrm{d}x,$$

the solution (P3.13) leads to

$$\Delta(z, y) = \int_z^y f^2(x)\,\mathrm{d}x - [\mathbf{F}(y) - \mathbf{F}(z)]^{\mathrm{T}}[\boldsymbol{\sigma}(y) - \boldsymbol{\sigma}(z)]^{-1}[\mathbf{F}(y) - \mathbf{F}(z)]. \qquad \text{(P3.14)}$$

### *P3.4.2 The Use of Legendre Polynomials*

Rather than dealing directly with the final polynomials, as above, let us decompose $f^*(x)$ in terms of Legendre polynomials, normalized to the interval $[z, y]$:

$$f^*(x) = \sum_{i=1}^{k} \beta^i P_i(t),$$

where

$$t = (2x - y - z)/(y - z), \qquad \text{(P3.15)}$$

since $x = z$ corresponds to $t = -1$, and $x = y$ to $t = 1$. Here $P_i(t)$ is the Legendre polynomial of degree $i$. Using the properties of the Legendre polynomials, the $\hat{\beta}^i$ are given by

$$\hat{\beta}^i = \frac{2i+1}{2} \int_{-1}^{1} F\left[\frac{(y-z)t + (y+z)}{2}\right] P_i(t)\,\mathrm{d}t. \qquad \text{(P3.16)}$$

From (P3.16) it can be seen that $k$ does not affect the values of the $\hat{\beta}^i$, on account of the properties of the Legendre polynomials. Then

$$\begin{aligned}
\Delta(z, y) &= \int_z^y \left[ f(x) - \sum_{i=1}^{k} \hat{\beta}^i P_i(t) \right]^2 \mathrm{d}x \\
&= \int_z^y f^2(x)\, \mathrm{d}x - \sum_i \int_z^y \hat{\beta}^i f(x) P_i(t)\, \mathrm{d}x \\
&\quad + \sum_i \sum_j \int_z^y \hat{\beta}^i \hat{\beta}^j P_i(t) P_j(t)\, \mathrm{d}x \\
&= \int_z^y f^2(x)\, \mathrm{d}x - \sum_i \frac{(y-z)}{2} \hat{\beta}^i \int_{-1}^{1} f(t) P_i(t)\, \mathrm{d}t \\
&\quad + \sum_i \sum_j \frac{y-z}{2} \hat{\beta}^i \hat{\beta}^j \int_{-1}^{1} P_i(t) P_j(t)\, \mathrm{d}t \\
&= \int_z^y f^2(x)\, \mathrm{d}x - \frac{(y-z)}{2} \sum_i \frac{2}{2i+1} \hat{\beta}^{i2},
\end{aligned}$$

or, finally,

$$\Delta(z, y) = \int_z^y f^2(x)\, \mathrm{d}x - (y-z) \sum_{i=1}^{k} \frac{\hat{\beta}^{i2}}{2i+1}. \tag{P3.17}$$

This formulation of the problem avoids inversion of a matrix. However, it does not make any simple use of the functions $\sigma_i(\ )$ which can be pre-calculated.

# Problem 4 | Regulation of Angular Position with Minimization of a Quadratic Criterion

## P4.1 Statement of the Problem

Let us consider a system which can be rotated by a motor, with the transfer function relating the angular position of the rotating system to the control $u$ applied to the motor being

$$F(p) = 1/[p(1 + p)] .$$

The control motor and the rotating system are driven in real time by a computer, which uses a control interval $T$, in the arrangement shown in Fig. P4.1. The digital-to-analog converter will be taken as a zero-order

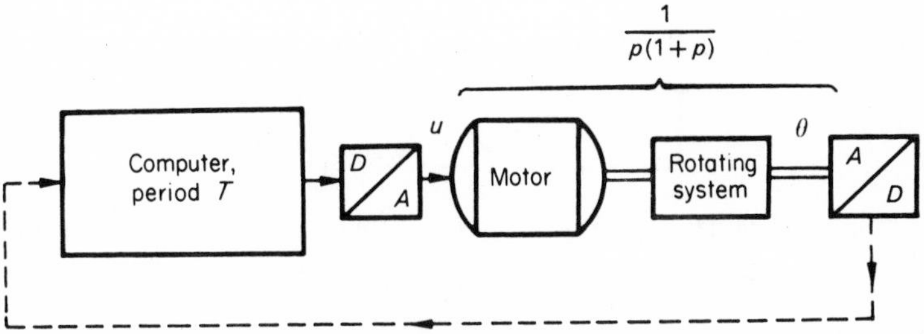


FIG. P4.1. The system to be controlled.

hold, and the angle coder will be assumed perfect, with no quantization error.

We wish to find the optimal control law, which regulates the angular position to some given value $\theta_f$ in some given time, while minimizing a quadratic criterion of the type $\sum u_n^2$, which might measure the energy consumed during the regulation time. We shall first establish a discrete model of the process, and determine the minimum-time control and the corresponding value attained by the cost criterion. We shall then establish directly the optimal recurrence relation, which leads to the solution of the problem of interest, and present the flow chart for the calculations involved and for the controller. For some particular $T$, we shall then calculate the form of the optimal trajectory. Finally, we shall determine the form of the calculations necessary to solve the problem of minimum-time regulation, with attained time $\hat{N}$, with the total energy consumption being bounded as:

$$\sum_{i=1}^{\hat{N}} u_i^2 \leq E.$$

## P4.2 Formulation of the Equations

Using the given transfer function, the controlled system obeys

$$\ddot{\theta} + \dot{\theta} = u(t). \tag{P4.1}$$

Using as components of the state vector

$$x_1 = \theta, \qquad x_2 = \dot{\theta},$$

this becomes

$$\dot{\boldsymbol{x}} = \begin{bmatrix} 0 & 1 \\ 0 & -1 \end{bmatrix} \boldsymbol{x} + \begin{bmatrix} 0 \\ 1 \end{bmatrix} u(t). \tag{P4.2}$$

Applying the Laplace transformation to (P4.2) results in

$$\begin{bmatrix} p & -1 \\ 0 & p+1 \end{bmatrix} \boldsymbol{X}(p) = \boldsymbol{x}_0 + \begin{bmatrix} 0 \\ 1 \end{bmatrix} u(p).$$

Assuming that $u(t) = u_0$ on $0 < t < T$, so that over the interval of interest

$$u(p) = u_0(1/p),$$

and using

$$\begin{bmatrix} p & -1 \\ 0 & p+1 \end{bmatrix}^{-1} = \frac{1}{p(p+1)} \begin{bmatrix} p+1 & 1 \\ 0 & p \end{bmatrix} = \begin{bmatrix} \dfrac{1}{p} & \dfrac{1}{p(p+1)} \\ 0 & \dfrac{1}{p+1} \end{bmatrix},$$

leads to

$$X(p) = \begin{bmatrix} \dfrac{1}{p} & \dfrac{1}{p(p+1)} \\ 0 & \dfrac{1}{p+1} \end{bmatrix} x_0 + \begin{bmatrix} \dfrac{1}{p^2(p+1)} \\ \dfrac{1}{p(p+1)} \end{bmatrix} u_0 .$$

But

$$1/p(p+1) = 1/p - 1/(p+1) \Rightarrow 1 - e^{-t},$$

$$1/p^2(p+1) = 1/p^2 - 1/p + 1/(p+1) \Rightarrow t - 1 + e^{-t},$$

so that

$$x(t) = \begin{bmatrix} 1 & 1 - e^{-t} \\ 0 & e^{-t} \end{bmatrix} x_0 + \begin{bmatrix} t - 1 + e^{-t} \\ 1 - e^{-t} \end{bmatrix} u_0, \qquad \text{(P4.3)}$$

describes the behavior of the system over one sampling interval. This finally leads to the desired discrete model,

$$x^+ = \begin{bmatrix} 1 & 1 - e^{-T} \\ 0 & e^{-T} \end{bmatrix} x + \begin{bmatrix} T - (1 - e^{-T}) \\ (1 - e^{-T}) \end{bmatrix} u = \mathbf{A}x + Bu . \qquad \text{(P4.4)}$$

## P4.3 Minimum-Time Control

Let us suppose that the required final state is the origin. Writing the system recursion (P4.4) two times,

$$x_1 = \mathbf{A}x_0 + Bu_0,$$

$$x_2 = \mathbf{A}x_1 + Bu_1 = \mathbf{A}^2 x_0 + \mathbf{A}Bu_0 + Bu_1 = 0 .$$

The matrix $\mathbf{A}$ is invertible, so that we can proceed as

$$\mathbf{A}^{-1}Bu_0 + \mathbf{A}^{-2}Bu_1 = -x_0,$$

$$\begin{bmatrix} u_0 \\ u_1 \end{bmatrix} = -[\mathbf{A}^{-1}B, \mathbf{A}^{-2}B]^{-1} x_0 = -\mathbf{M}X_0,$$

$$\mathbf{A}^{-1} = e^T \begin{bmatrix} e^{-T} & e^{-T} - 1 \\ 0 & 1 \end{bmatrix} = \begin{bmatrix} 1 & 1 - e^T \\ 0 & e^T \end{bmatrix},$$

$$\mathbf{A}^{-1}B = \begin{bmatrix} 1 & 1 - e^T \\ 0 & e^T \end{bmatrix} \begin{bmatrix} T - (1 - e^{-T}) \\ 1 - e^{-T} \end{bmatrix} = \begin{bmatrix} T - (e^T - 1) \\ e^T - 1 \end{bmatrix},$$

$$\mathbf{A}^{-1}(\mathbf{A}^{-1}B) = \begin{bmatrix} 1 & 1 - e^T \\ 0 & e^T \end{bmatrix} \begin{bmatrix} T - (e^T - 1) \\ e^T - 1 \end{bmatrix} = \begin{bmatrix} T - e^T(e^T - 1) \\ e^T(e^T - 1) \end{bmatrix},$$

$$[\mathbf{A}^{-1}B, \mathbf{A}^{-2}B] = \begin{bmatrix} T - (e^T - 1) & T - e^T(e^T - 1) \\ (e^T - 1) & e^T(e^T - 1) \end{bmatrix}.$$

The determinant of this last matrix is

$$\begin{aligned}\Delta &= Te^T(e^T - 1) - e^T(e^T - 1)^2 - T(e^T - 1) + e^T(e^T - 1)^2 \\ &= T(e^T - 1)^2 \neq 0\end{aligned}$$

so that the inverse exists, and it is thus possible to steer the system in two steps to any arbitrary state (the system is controllable). Specifically,

$$\mathbf{M} = [\mathbf{A}^{-1}\boldsymbol{B}, \mathbf{A}^{-2}\boldsymbol{B}]^{-1} = \begin{bmatrix} \dfrac{e^T}{T(e^T - 1)} & -\dfrac{T - e^T(e^T - 1)}{T(e^T - 1)^2} \\ -\dfrac{1}{T(e^T - 1)} & \dfrac{T - (e^T - 1)}{T(e^T - 1)^2} \end{bmatrix}, \qquad \text{(P4.5)}$$

so that the control to be applied is

$$\begin{bmatrix} u_0 \\ u_1 \end{bmatrix} = \begin{bmatrix} -\dfrac{e^T}{T(e^T - 1)} & \dfrac{T - e^T(e^T - 1)}{T(e^T - 1)^2} \\ \dfrac{1}{T(e^T - 1)} & \dfrac{(e^T - 1) - T}{T(e^T - 1)^2} \end{bmatrix} \boldsymbol{x}_0 = -\mathbf{M}\boldsymbol{x}_0 . \qquad \text{(P4.6)}$$

According to the properties of a minimum-time regulator [1, Vol. 1, Sect. 6.6.1], this law can be realized by applying, at each control time,

$$u = -\left[\frac{e^T}{T(e^T - 1)}, \quad \frac{e^T(e^T - 1) - T}{T(e^T - 1)^2}\right]\boldsymbol{x} . \qquad \text{(P4.7)}$$

This corresponds to the system shown in Fig. P4.2, in which the desired

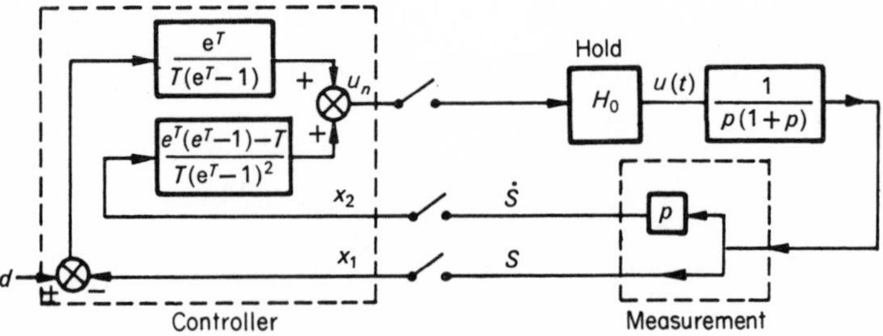


FIG. P4.2. The controlled system.

final position is shown as an input. Since without loss of generality this final position can be taken as the origin, the assumption that the final state is the origin is justified.

Letting $\mathscr{R}_2(\boldsymbol{x}_0)$ be the value of the criterion after two steps, using the optimal control, we have

$$\mathscr{R}_2(\boldsymbol{x}_0) = u_0{}^2 + u_1{}^2 = [u_0 \quad u_1]\begin{bmatrix} u_0 \\ u_1 \end{bmatrix} = \boldsymbol{x}_0{}^{\mathrm{T}}\mathbf{M}^{\mathrm{T}}\mathbf{M}\boldsymbol{x}_0\,,$$

or

$$\mathscr{R}_2(\boldsymbol{X}) = \boldsymbol{X}^{\mathrm{T}}\mathbf{Q}_2\,\boldsymbol{X}\,,$$

where

$\mathbf{Q}_2$

$$= \frac{1}{T^2(\mathrm{e}^T - 1)^4}\begin{bmatrix} -\mathrm{e}^T(\mathrm{e}^T - 1) & \mathrm{e}^T - 1 \\ T - \mathrm{e}^T(\mathrm{e}^T - 1) & (\mathrm{e}^T - 1) - T \end{bmatrix}\begin{bmatrix} -\mathrm{e}^T(\mathrm{e}^T - 1) & T - \mathrm{e}^T(\mathrm{e}^T - 1) \\ (\mathrm{e}^T - 1) & (\mathrm{e}^T - 1) - T \end{bmatrix}$$

$$= \begin{bmatrix} \dfrac{1 + \mathrm{e}^{2T}}{T^2(\mathrm{e}^T - 1)^2} & \dfrac{(\mathrm{e}^T - 1)(1 + \mathrm{e}^{2T}) - T(1 + \mathrm{e}^T)}{T^2(\mathrm{e}^T - 1)^3} \\[2ex] \dfrac{(\mathrm{e}^T - 1)(1 + \mathrm{e}^{2T}) - T(1 + \mathrm{e}^T)}{T^2(\mathrm{e}^T - 1)^3} & \dfrac{2T^2 - 2T(\mathrm{e}^{2T} - 1) + (\mathrm{e}^T - 1)^2(1 + \mathrm{e}^{2T})}{T^2(\mathrm{e}^T - 1)^4} \end{bmatrix} \tag{P4.8}$$

## P4.4 The Optimal Recurrence Equation

The system being stationary, the optimal return depends only on the state $\boldsymbol{X}$ and the number $m$ of control cycles remaining until the stopping time. It is thus a function $\mathscr{R}_m(\boldsymbol{X})$, satisfying the optimal recurrence relation

$$\mathscr{R}_m(\boldsymbol{X}) = \min_u \{u^2 + \mathscr{R}_{m-1}(\mathbf{A}\boldsymbol{X} + \boldsymbol{B}u)\}\,. \tag{P4.9}$$

Equation (P4.9) is to be initialized at $m = 2$ for which, as we have seen, the condition $\boldsymbol{X}_{\mathrm{f}} = 0$ completely determines $u_0$ and $u_1$. Further, $\mathscr{R}_2(\boldsymbol{X})$ is a quadratic form $\boldsymbol{X}^T\mathbf{Q}_2\boldsymbol{X}$. Let us demonstrate that this is generally the case for $\mathscr{R}_m$ by assuming, as an inductive hypothesis, that $\mathscr{R}_{m-1}(\boldsymbol{X}) = \boldsymbol{X}^T\mathbf{Q}_{m-1}\boldsymbol{X}$.

The term in braces in (P4.9) can be written, with this assumption, as

$$\begin{aligned}\{\,\} &= u^2 + (\mathbf{A}\boldsymbol{X} + \boldsymbol{B}u)^{\mathrm{T}}\mathbf{Q}_{m-1}(\mathbf{A}\boldsymbol{X} + \boldsymbol{B}u) \\ &= u^2(1 + \boldsymbol{B}^{\mathrm{T}}\mathbf{Q}_{m-1}\boldsymbol{B}) + 2u\boldsymbol{B}^{\mathrm{T}}\mathbf{Q}_{m-1}\mathbf{A}\boldsymbol{X} + \boldsymbol{X}^{\mathrm{T}}\mathbf{A}^{\mathrm{T}}\mathbf{Q}_{m-1}\mathbf{A}\boldsymbol{X}\,.\end{aligned}$$

The minimum is attained for

$$\hat{u} = -\frac{1}{1 + \boldsymbol{B}^{\mathrm{T}}\mathbf{Q}_{m-1}\boldsymbol{B}}\boldsymbol{B}^{\mathrm{T}}\mathbf{Q}_{m-1}\mathbf{A}\boldsymbol{X} = -\boldsymbol{L}_m{}^{\mathrm{T}}\boldsymbol{X}\,, \tag{P4.10}$$

where we define

$$\boldsymbol{L}_m = \frac{1}{1 + \boldsymbol{B}^{\mathrm{T}}\mathbf{Q}_{m-1}\boldsymbol{B}}\,\mathbf{A}^{\mathrm{T}}\mathbf{Q}_{m-1}\boldsymbol{B}. \tag{P4.11}$$

Substituting this $\hat{u}$ into (P4.9), there indeed results a quadratic form with matrix $\mathbf{Q}_m$ given by

$$\mathbf{Q}_m = \mathbf{A}^{\mathrm{T}}\mathbf{Q}_{m-1}\mathbf{A} - (1 + \boldsymbol{B}^{\mathrm{T}}\mathbf{Q}_{m-1}\boldsymbol{B})\,\boldsymbol{L}_m\,\boldsymbol{L}_m^{\mathrm{T}}. \tag{P4.12}$$

This relation is to be initialized with $\mathbf{Q}_2$ as in (P4.8).

Relations (P4.11) and (P4.12) constitute a system of recursions which determine the linear optimal control law $\boldsymbol{L}_k$ and the corresponding matrices appearing in the optimal return function:

$$\begin{aligned} V_k &= \boldsymbol{B}^{\mathrm{T}}\mathbf{Q}_{k-1}\boldsymbol{B} + 1, \\ \boldsymbol{L}_k^{\mathrm{T}} &= V_k^{-1}\boldsymbol{B}^{\mathrm{T}}\mathbf{Q}_{k-1}\mathbf{A}, \\ \mathbf{Q}_k &= \mathbf{A}^{\mathrm{T}}\mathbf{Q}_{k-1}\mathbf{A} - V_k\,\boldsymbol{L}_k\,\boldsymbol{L}_k^{\mathrm{T}}. \end{aligned} \tag{P4.13}$$

These relations agree with those established in the general case, formulas (P4.5) and (P4.6) of Chapter 8.

Let us expand the system (P4.13) in scalar form, defining

$$\mathbf{A} = \begin{bmatrix} 1 & A_1 \\ 0 & A_2 \end{bmatrix}, \qquad \boldsymbol{B} = \begin{bmatrix} B_1 \\ B_2 \end{bmatrix},$$

$$\mathbf{Q} = \begin{bmatrix} Q_1 & Q_3 \\ Q_3 & Q_2 \end{bmatrix}, \qquad \boldsymbol{L} = \begin{bmatrix} G \\ H \end{bmatrix}.$$

Using these in (P4.13)

$$V = [B_1 \quad B_2]\begin{bmatrix} Q_1^- & Q_3^- \\ Q_3^- & Q_2^- \end{bmatrix}\begin{bmatrix} B_1 \\ B_2 \end{bmatrix} + 1,$$

$$[G \quad H] = \frac{1}{V}[B_1 \quad B_2]\begin{bmatrix} Q_1^- & Q_3^- \\ Q_3^- & Q_2^- \end{bmatrix}\begin{bmatrix} 1 & A_1 \\ 0 & A_2 \end{bmatrix},$$

$$\begin{bmatrix} Q_1 & Q_3 \\ Q_3 & Q_2 \end{bmatrix} = \begin{bmatrix} 1 & 0 \\ A_1 & A_2 \end{bmatrix}\begin{bmatrix} Q_1^- & Q_3^- \\ Q_3^- & Q_2^- \end{bmatrix}\begin{bmatrix} 1 & A_1 \\ 0 & A_2 \end{bmatrix} - \begin{bmatrix} G \\ H \end{bmatrix} V [G \quad H],$$

or

$$\begin{aligned} V &= 1 + B_1^2 Q_1^- + 2B_1 B_2 Q_3^- + B_2^2 Q_2^-, \\ G &= (B_1 Q_1^- + B_2 Q_3^-)/V, \\ H &= [B_1(A_1 Q_1^- + A_2 Q_3^-) + B_2(A_1 Q_3^- + A_2 Q_2^-)]/V, \\ Q_1 &= Q_1^- - G^2 V, \\ Q_2 &= A_1 Q_1^- + A_2 Q_3^- - GHV, \\ Q_3 &= A_1(A_1 Q_1^- + A_2 Q_3^-) + A_2(A_1 Q_3^- + A_2 Q_2^-) - H^2 V. \end{aligned} \tag{P4.14}$$

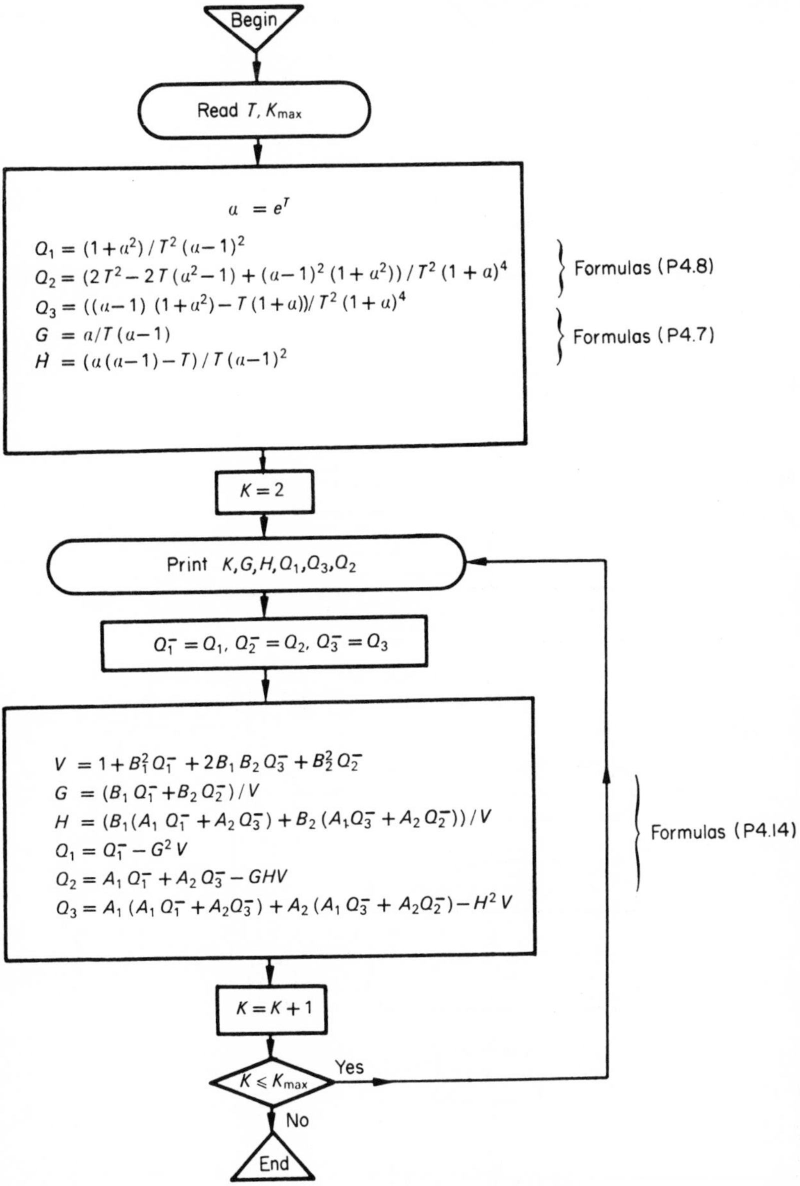


FIG. P4.3. Synthesis of the controller.

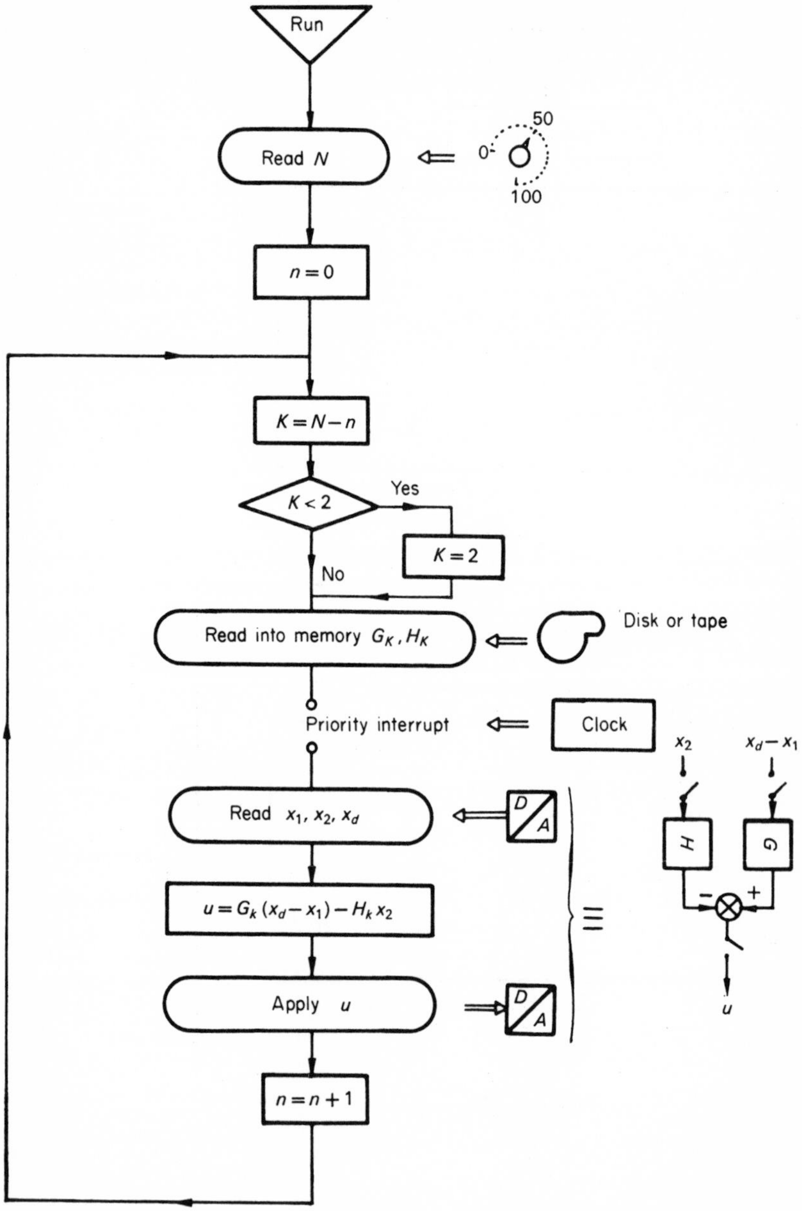


FIG. P4.4. The optimal controller.

These formulas, along with (P4.7) and (P4.8), allow computation of the minimum-time control law, and of $\mathbf{Q}_2$, according to the flow chart in Fig. P4.3. The flow chart of Fig. P4.4 shows the optimal controller.

## P4.5 Example of an Optimal Trajectory

For $T = 0.2$, the above calculations result in

$$\mathbf{A} = \begin{bmatrix} 1 & 0.18127 \\ 0 & 0.81873 \end{bmatrix}, \qquad B = \begin{bmatrix} 0.01873 \\ 0.18127 \end{bmatrix},$$

$$\mathbf{M} = \begin{bmatrix} -27.582 & -7.183 \\ 22.582 & 2.1829 \end{bmatrix}, \qquad \mathbf{Q}_2 = \begin{bmatrix} 1270.8 & 247.4 \\ 247.4 & 56.3 \end{bmatrix},$$

leading to the optimal control law shown in Table P4.1. Use of this optimal

**Table P4.1**

| $N$ | 1 | 2 | 3 | 4 | 5 | 6 | 7 |
|---|---|---|---|---|---|---|---|
| $G$ | 27.582 | 27.582 | 13.792 | 8.286 | 5.538 | 3.971 | 2.994 |
| $H$ | 7.183 | 7.183 | 4.943 | 3.693 | 2.897 | 2.348 | 1.949 |

| $N$ | 8 | 9 | 10 | 11 | 12 | 13 | 14 | 15 |
|---|---|---|---|---|---|---|---|---|
| $G$ | 2.343 | 1.889 | 1.559 | 1.311 | 1.121 | 0.972 | 0.853 | 0.756 |
| $H$ | 1.648 | 1.413 | 1.227 | 1.086 | 0.952 | 0.849 | 0.763 | 0.689 |

control leads to Fig. P4.5, starting from $x_1 = 1$, $x_2 = 0$, and to Fig. P4.6 starting from $x_1 = 0$, $x_2 = 1$.

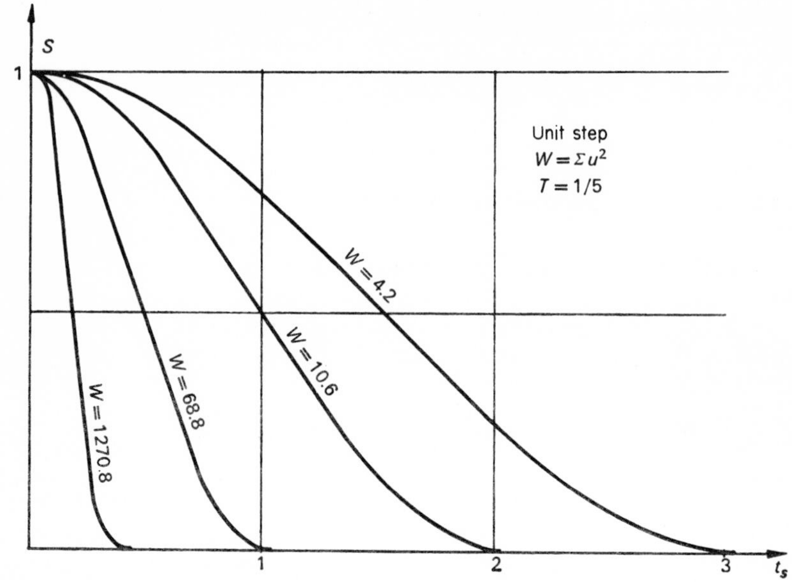


FIG. P4.5. Regulation of the controlled system to remove angle error.

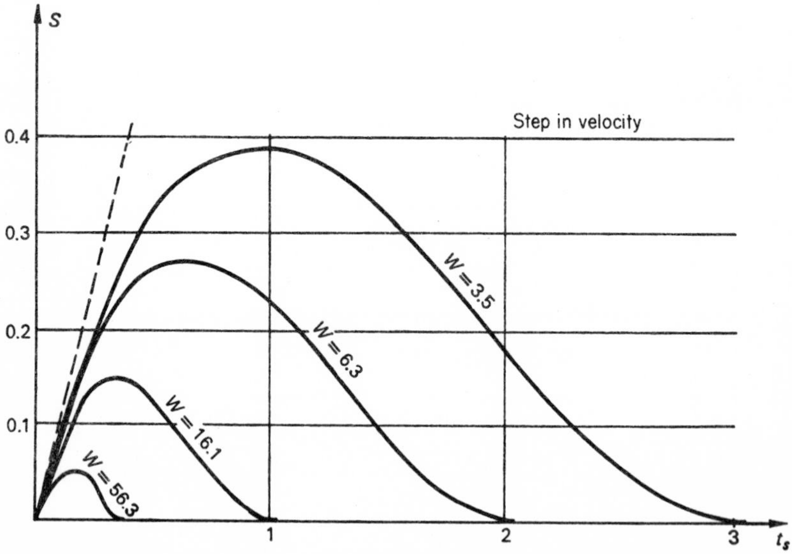


FIG. P4.6. Regulation of the controlled system to remove rate error.

## P4.6 Minimum-Time Control with Restricted Energy

Suppose the initial state is $\boldsymbol{x}_0$, and let

$$\mathscr{R}_N(\boldsymbol{x}_0) = \boldsymbol{x}_0{}^{\mathrm{T}}\mathbf{Q}_N \boldsymbol{x}_0 .$$

The return $\mathscr{R}_N(\boldsymbol{x}_0)$ is a monotonically decreasing function of $N$, and we wish to find that value of $N$ for which there first occurs

$$\mathscr{R}_N(\boldsymbol{x}_0) = \boldsymbol{x}_0{}^{\mathrm{T}}\mathbf{Q}_N \boldsymbol{x}_0 \leq E^2 .$$

This leads at once to the flow chart shown in Fig. P4.7. Note that in

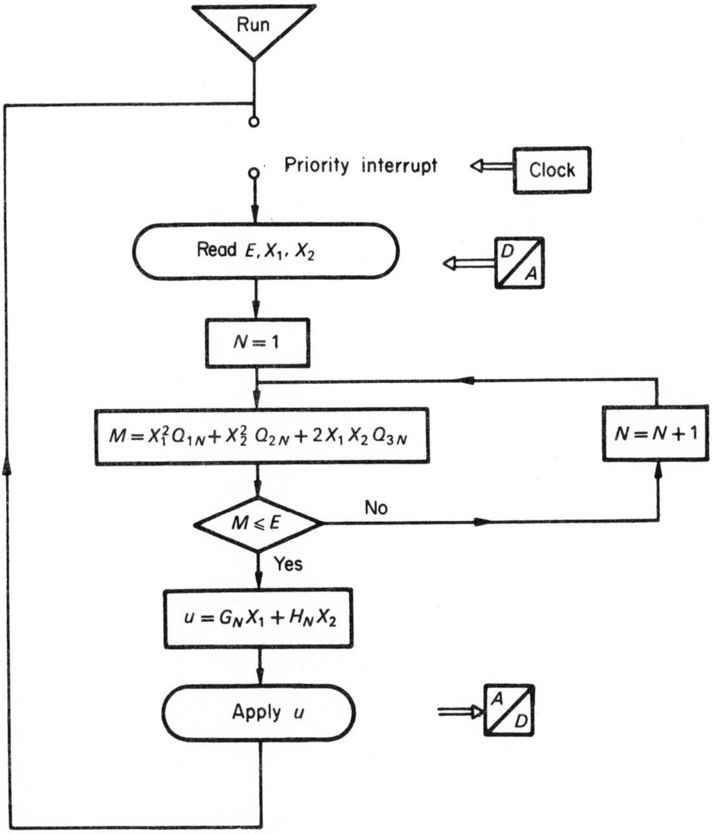


FIG. P4.7. Optimal control with bounded energy.

order that the delay between reading $(x_1, x_2)$ and applying the corresponding control be small, the quantities $G_n$, $H_n$, $Q_{1n}$, $Q_{2n}$, and $Q_{3n}$ should be in the fast memory of the computer.

# Problem 5 | Control of a Stochastic System

## P5.1 Statement of the Problem

Let us consider the first-order discrete process diagrammed in Fig. P5.1.

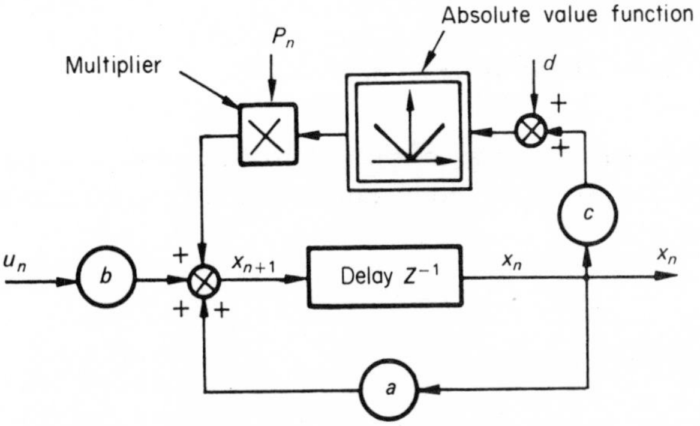


FIG. P5.1. The first-order discrete process to be controlled.

Two external signals act on this system. The first is a constant, and measurable, bias input, $d$. The second is a random perturbation sequence $P_n$, which is not available for measurement. The value of $P_n$ at each process step is independent of its values at the previous steps, and its distribution function is known, as shown in Fig. P5.2.

In order to determine the optimal control law for this system, we shall first formulate a stochastic model of it. The objective of control is to

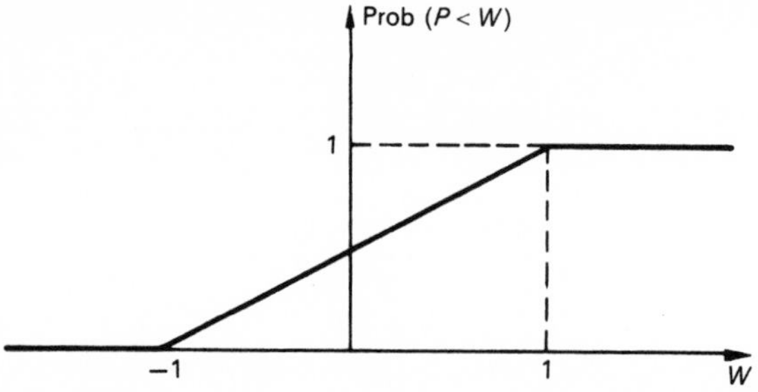


FIG. P5.2. Distribution function of the random noise in the system.

minimize the mathematical expectation of the absolute value of the state variable at a final time $N$ specified in advance. We shall formulate and solve the optimal recurrence equation satisfied by the optimal return function and thereby determine the optimal control policy. The flow chart of the optimal controller will be given. We shall also determine a controller to minimize the mathematical expectation of the time at the end of which the state has absolute value less than $\varepsilon$. The implicit equation which the optimal return must satisfy in this case will be established.

## P5.2 Stochastic Model of the System

The mathematical model of this stochastic system is taken to be the distribution function of the future state, for a given present state and a given control signal:

$$G(y, x_n, u_n) = \text{prob}\{x_{n+1} \leq y \,|\, x_n, u_n\}. \tag{P5.1}$$

Using Fig. P5.1, we have

$$x_{n+1} = ax_n + bu_n + |cx_n + d|\,P_n, \tag{P5.2}$$

from which

$$\begin{aligned} G(y, x_n, u_n) &= \text{prob}\left\{P_n \leq \frac{y-(ax_n+bu_n)}{|cx_n+d|}\,\middle|\,x_n, u_n\right\} \\ &= F\left[\frac{y-(ax_n+bu_n)}{|cx_n+d|}\right], \end{aligned} \tag{P5.3}$$

where $F[\,]$ is the distribution function shown in Fig. P5.3.

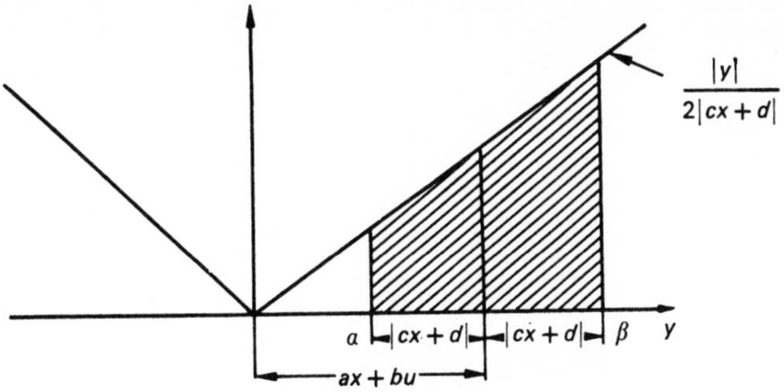


FIG. P5.3. The integral to be minimized.

For convenience in future calculations, let us determine the probability density corresponding to the distribution $G(y, x_n, u_n)$:

$$g(y) = \frac{\partial}{\partial y} G(y, x_n, u_n) = \begin{cases} 0, & y \notin I, \\ \dfrac{1}{2|cx_n + d|}, & y \in I, \end{cases} \tag{P5.4}$$

where $I$ is the interval

$$[ax_n + bu_n - |cx_n + d|, \qquad ax_n + bu_n + |cx_n + d|]. \tag{P5.5}$$

## P5.3 Optimal Recurrence Equation with Terminal Cost

The optimal return function depends on the initial state $X$ and on the number $m$ of control intervals separating the present time $n$ and the terminal time $N = n + m$. It is thus a function of the form $\hat{R}_m(X)$, and satisfies the recurrence, obtained by direct application of the principle of optimality,

$$\hat{R}_m(x_n) = \min_u \mathrm{E}\{\hat{R}_{m-1}(x_{n+1})\}, \tag{P5.6}$$

in which the mathematical expectation is relative to the transition $x_n \rightarrow x_{n+1}$, with $x_n$ and $u_n$ known.

The expectation in (P5.6) can be evaluated, using $g(y)$, the probability density of $x_{n+1}$, to obtain

$$\hat{R}_m(x) = \min_u \int_{-\infty}^{+\infty} \hat{R}_{m-1}(y) g(y)\, dy$$

$$= \min_u \int_{\alpha}^{\beta} \hat{R}_{m-1}(y) \frac{1}{2|cx + d|}\, dy, \qquad \text{(P5.7)}$$

$$\alpha = ax + bu - |cx + d|,$$

$$\beta = ax + bu + |cx + d|.$$

## P5.4 Solution of the Optimality Equation

Let us begin the solution with $m = 1$. In this case, $\hat{R}_{m-1}(x)$ is just $\hat{R}_0(x)$, the expected value of $|x|$ given $x$, which is simply

$$\hat{R}_0(x) = |x|. \qquad \text{(P5.8)}$$

Relation (P5.7) then yields

$$\hat{R}_1(x) = \min_u \int_{\alpha}^{\beta} \frac{|y|}{2|cx + d|}\, dy.$$

The integral in this last relation can be analyzed using the sketch in Fig. P5.3. It is easy to see that the minimum value of the integral, the shaded area in the figure, occurs for

$$ax + b\hat{u} = 0 \rightarrow \hat{u} = -ax/b. \qquad \text{(P5.9)}$$

This corresponds to an attained minimum

$$\hat{R}_1(x) = 2 \times \frac{1}{2}\left(|cx + d| \frac{|cx + d|}{2|cx + d|}\right),$$

or

$$\hat{R}_1(x) = \frac{|cx + d|}{2}. \qquad \text{(P5.10)}$$

Substituting the value of $\hat{R}_1(x)$ in (P5.10) into the recursion relation (P5.7), with $m = 2$, results in

$$\hat{R}_2(x) = \min_u \int_{\alpha}^{\beta} \frac{|cy + d|}{2} \frac{1}{2|cx + d|}\, dy.$$

This integral can be represented as the shaded area in Fig. P5.4, and is minimum for

$$ax + b\hat{u} = -d/c,$$

$$\hat{u} = -ax/b - d/cb. \tag{P5.11}$$

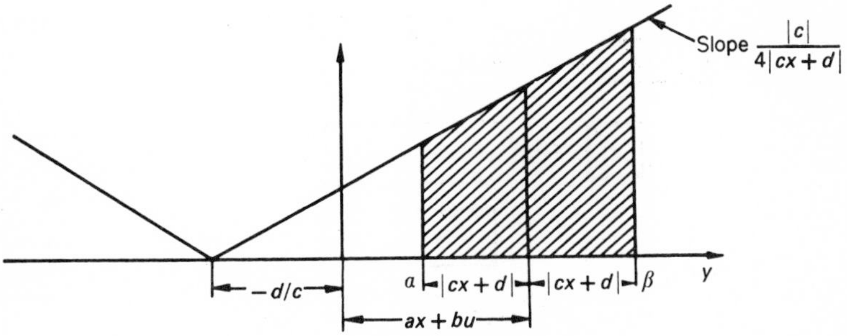


FIG. P5.4. The integral to be minimized at the second step.

The attained minimum value is

$$\hat{R}_2(x) = 2 \times \frac{1}{2}\left\{|cx+d| \frac{|cx+d|\,|c|}{4|cx+d|}\right\} = \frac{|c|\,|cx+d|}{4}$$

$$= \frac{|c|}{2}\hat{R}_1(x). \tag{P5.12}$$

Since $\hat{R}_2(x)$ differs from $\hat{R}_1(x)$ only by a fixed multiplier $|c|/2$, the subsequent returns can be calculated by successive applications of that multiplier. Thus

$$\hat{R}_m(x) = \left|\frac{c}{2}\right|^{m-1} \frac{|cx+d|}{2}, \tag{P5.13}$$

with the optimal control being

$$\hat{u}_m = \begin{cases} -ax/b, & m = 1, \\ -ax/b - d/cb, & m \neq 1. \end{cases} \tag{P5.14}$$

The control law (P5.14) corresponds to the flow chart shown in Fig. P5.5. Note that the measurable quantity $d$ is entered at each cycle, so that changes in $d$ during system operation can be accommodated. The control policy is optimal after a change in $d$, if it is assumed that it is the last to occur.

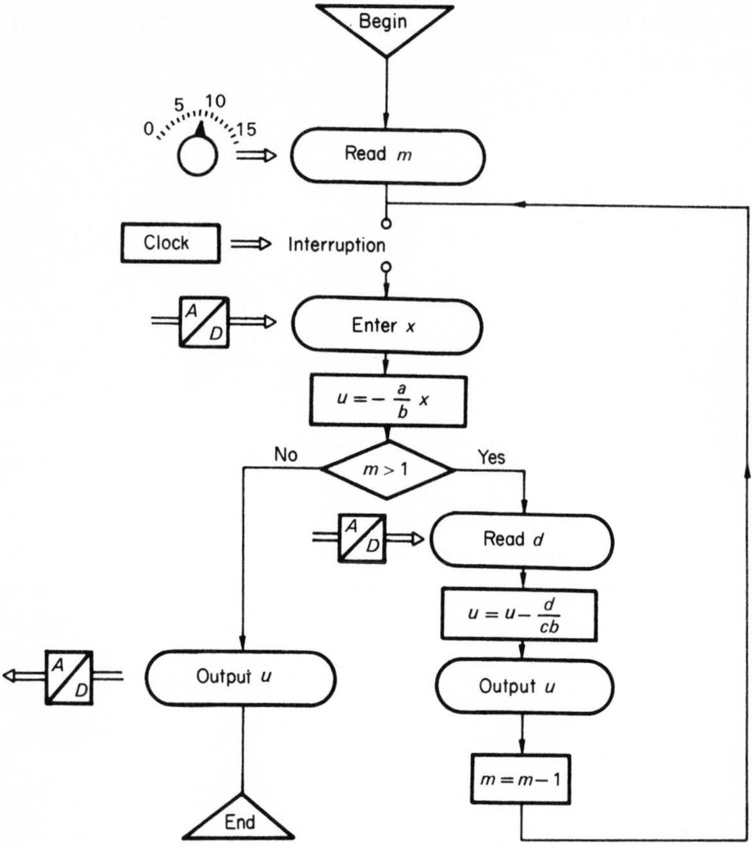


FIG. P5.5. The optimal controller.

## P5.5 Minimum-Time Criterion

The mathematical expectation of the time at the end of which the state will be less in absolute value than $\varepsilon$, for a given control policy, depends only on the initial state $x$ for a stationary process. Then let $T(x)$ be the return function corresponding to an optimal policy. Using the principle of optimality,

$$\begin{aligned} T(x) &= 1 + \min_u \mathrm{E}[T(x^+)], \qquad |x| > \varepsilon, \\ T(x) &= 0, \qquad |x| < \varepsilon. \end{aligned} \tag{P5.15}$$

Calculating the expectation using $g(y)$ results in

$$T(x) = 1 + \min_u \int_{ax+bu-|cx+d|}^{ax+bu+|cx+d|} T(y) \frac{1}{2|cx+d|} \, dy, \qquad |x| > \varepsilon,$$
$$T(x) = 0, \qquad |x| < \varepsilon. \tag{P5.16}$$

Relation (P5.16) is an implicit functional equation which, in general, can be solved only by successive approximations. We shall not proceed further with the solution here.

# Problem 6 Minimum-Time Depth Change of a Submersible Vehicle

## P6.1 Statement of the Problem

The vertical motion of a submersible vehicle, having zero horizontal velocity, can be represented by

$$M\ddot{z} + k\dot{z}|\dot{z}| = F,$$

where $z$ is the depth of the vehicle, $M$ the effective mass (about twice the actual mass, taking account of the volume of water displaced), $k$ the hydrodynamic drag coefficient, and $F$ the vertical force exerted by the submersing system of the vehicle. Starting from the initial state $z = z_i$, $\dot{z} = 0$, it is desired to carry the vehicle to the final state $z = z_f$, $\dot{z} = 0$, in the shortest possible time. The force $F$ is required to be such that $F_1 \leq F \leq F_2$.

In order that the operation be carried out without overshoot of the desired final depth, the motion is, in addition, required to be such that $\dot{z}$ is always of the same sign. If $z_f \geq z_i$, for example, this requires that $\dot{z} \geq 0$.

It is assumed that the control law will be generated using a digital computer, and that the optimal control force $F$ will be calculated during the operation, in terms of the depth deviations $x_n$ defined by

$$z = z_f - x_n,$$

with $z_0 = 0$ and

$$x_n - x_{n-1} = \Delta_n.$$

This value of $F$ will be held until reaching the following value $x_{n-1}$. The scheme is diagrammed in Fig. P6.1.

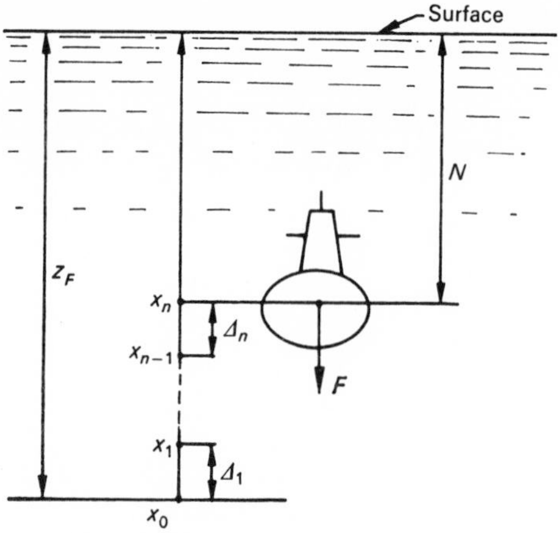


FIG. P6.1. The problem geometry.

Considering the condition imposed on $\dot{z}$, and the resulting control characteristics, the independent variable is taken to be the depth $z$, rather than time. The recursive mathematical model describing the process evolution can then be determined, for passage between two particular depths. The expression for the time required to carry out this depth change can then be found, and it is this which is the criterion function, since we are dealing with a minimum-time problem. The properties of the Legendre polynomials are used in this calculation for approximating integrals. The optimal recurrence equation is then written.

Finally, minimum-time submersion is considered with two types of constraints. First, $\int |F|\, dt$ is constrained, since this is approximately the energy consumed during the maneuver if the force $F$ is obtained with a screw device. Finally, a constraint on $\sum |F_n - F_{n-1}|$ is considered, since this is approximately the amount of compressed air used up in the maneuver if $F$ results from the Archimedean thrust obtained by moving water from outside the vehicle into and out of two internal tanks, as shown in Fig. P6.2. One tank is maintained at a pressure above the exterior pressure allowing negative $\Delta F$, and the other is maintained below external pressure corresponding to positive $\Delta F$.

## P6.2 Formulation of the Model

Letting

$$y = \dot{z}^2, \qquad x = z_f - z \tag{P6.1}$$

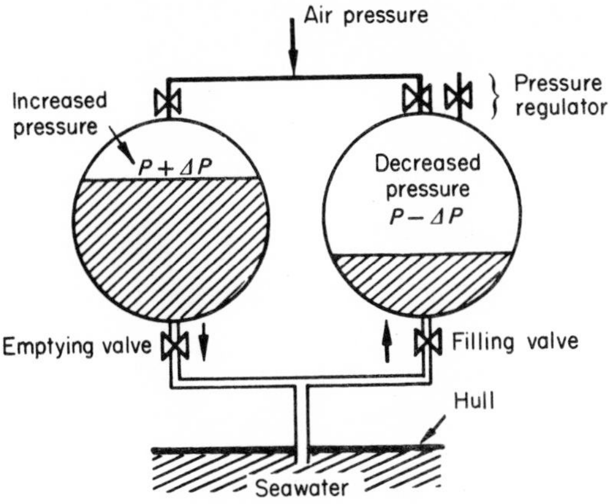


FIG. P6.2. The submersion system of the vehicle.

results in

$$\begin{aligned} dy &= 2\,\dot{z}\ddot{z}\,dt \\ dx &= -\dot{z}\,dt \end{aligned} \qquad dy/dx = -2\,\ddot{z}\,. \tag{P6.2}$$

In the case that $\dot{z}$ is always positive, the equation of motion of the vehicle can be written

$$-(M/2)\,(dy/dx) + ky = F\,.$$

Letting $\omega = 2k/M$ and $u = F/k$ results in the first-order linear equation

$$-(dy/dx) + \omega y = \omega u\,. \tag{P6.3}$$

Integration of this equation yields the relation $\dot{z}^2 = f(z)$.

In the case that $x$ changes from $x = x_n$ to $x_{n-1} = x_n - \Delta_n$, under the action of the control $u_n$, integration of (P6.3) yields

$$y_{n-1} = e^{-\omega\Delta_n} y_n + u_n(1 - e^{-\omega\Delta_n})\,, \tag{P6.4}$$

which is a linear stationary recurrence relation, in the case that the $\Delta_n$ are equal. It develops in the reverse direction, since the index 0 is used for the final state.

## P6.3 Formulation of the Criterion Function

Since

$$y = [dx/dt]^2\,,$$

we have

$$dt = dx/\sqrt{y}\,,$$

which leads to

$$t_{n-1} - t_n = \int_0^{\Delta_n} \frac{d\theta}{[y(\theta)]^{1/2}}, \tag{P6.5}$$

where $y(\theta)$ is obtained by integrating (P6.3) from $x_n$ to $x_n - \theta$:

$$y(\theta) = e^{-\omega\theta} y_n + (1 - e^{-\omega\theta}) u_n .$$

The criterion function is thus given in terms of

$$r_n(y_n, u_n) = t_{n-1} - t_n = \int_0^{\Delta_n} \frac{d\theta}{[e^{-\omega\theta}(y_n - u_n) + u_n]^{1/2}}. \tag{P6.6}$$

Unfortunately, literal calculation of this integral is not possible. We shall hence make use of the approximate formula, involving the Legendre polynomials,

$$\int_{-1}^{1} f(x)\, dx \simeq \sum_{i=1}^{k} H_i f(x_i) .$$

To make use of this formula, it is first necessary to make a change of variable to carry the interval $[0, \Delta_n]$ into $[-1, 1]$. Thus we define $v$ by

$$\theta = \Delta_n(v + 1)/2,$$

which leads to

$$r_n(y_n, u_n) = \int_{-1}^{1} \varphi(v, y_n, u_n)\, dv$$

with

$$\varphi(v, y_n, u_n) = \frac{\Delta_n}{2[e^{-(\omega\Delta/2)v}(y_n - u_n) + e^{-\omega\Delta_n/2}(y_n - u_n) + u_n]^{1/2}}.$$

The approximate formula (4.11) of Chapter 4 then yields

$$r_n(y_n, u_n) \simeq \sum_{1}^{k} H_i \varphi(v_i, y_n, u_n), \tag{P6.7}$$

in which the values $H_i$ and $v_i$ can be calculated in advance, depending on the degree $k$ chosen for the approximation.

## P6.4 The Recurrence Equation

We now have to consider the following discrete problem: Starting from $x_N = z_f - z_i$, with $y_N = \dot{z}_i^2 = 0$, find the optimal $N$-step law which restores the system to $y_0 = 0$, where the system is described by (P6.4), the criterion function is (P6.6), and the constraint

$$F_1/k = U_m \leq u_n \leq U_M = F_2/k$$

is present.

To solve this problem, we shall consider the class of problems for which $N$ and $y_N$ are arbitrary. The optimal return (i.e., the optimal maneuver time) will be a function $\mathscr{R}_n(y)$ depending on the operation to be carried out ($n$), and on the initial vertical velocity ($y$). This function satisfies the optimal recurrence relation

$$\mathscr{R}_n(y) = \min_{U_m \leq u \leq U_M} \left\{ \sum_1^k H_i \varphi(v_i, y_n, u_n) + \mathscr{R}_{n-1}[\mathrm{e}^{-\omega\Delta_n} y + (1 - \mathrm{e}^{-\omega\Delta_n})u] \right\}. \tag{P6.8}$$

This equation is to be initialized with the return $\mathscr{R}_1(y)$. Since $u_1$ is determined uniquely by the final state as

$$u_1 = [1/(\mathrm{e}^{\omega\Delta_1} - 1)]y_1 , \tag{P6.9}$$

the return $\mathscr{R}_1(y)$ is just

$$\mathscr{R}_1(y) = \sum_i H_i \varphi(v_i, y_1, y_1/(\mathrm{e}^{\omega\Delta_1} - 1)) . \tag{P6.10}$$

This function is defined only on the controllable interval

$$(\mathrm{e}^{\omega\Delta_1} - 1)(F_1/k) < y_1 < (\mathrm{e}^{\omega\Delta_1} - 1)(F_2/k) . \tag{P6.11}$$

This formulation of the problem leads to a one-dimensional linear system, which simplifies the solution of the recurrence relation (P6.8).

## P6.5 Introduction of Constraints

### *P6.5.1 Constraint on Energy Consumption*

The instantaneous energy consumption $|F|\ \mathrm{d}t$ is

$$|F|\ \mathrm{d}t = k|u|\ \mathrm{d}t = k|u| \frac{\mathrm{d}x}{\sqrt{v}} .$$

The energy consumption during passage from $x_n$ to $x_{n-1}$ is then

$$c_n(y_n, u_n) = k\,|u_n|\,r_n(y_n, u_n). \tag{P6.12}$$

The total energy consumed up till passage through $x_0$ thus satisfies the recursion

$$\mathscr{C}_{n-1} = \mathscr{C}_n + c_n(y_n, u_n), \qquad \mathscr{C}_{\text{initial}} = 0. \tag{P6.13}$$

The optimal return now depends on the amount of energy already consumed, and thus the energy consumption should be considered part of the state vector, with (P6.4) and (P6.13) now being the system equations. The optimal return is then a function $\mathscr{R}_n(y_n, \mathscr{C}_n)$, and satisfies the recurrence relation

$$\mathscr{R}_n(y, \mathscr{C}) = \min_{u \in \Omega}\left\{\sum_{i=1}^{k} H_i\varphi(v_i, y, u) + \mathscr{R}_{n-1}\left[\mathrm{e}^{-\omega\Delta_n}y + (1 - \mathrm{e}^{-\omega\Delta_n})u,\right.\right.$$
$$\left.\left.\mathscr{C}_n + k|u_n|\sum_{i=1}^{k} H_i\varphi(v_i, y, u)\right]\right\}. \tag{P6.14}$$

The domain $\Omega$ consists of all $u$ such that

$$U_m < u < U_M,$$
$$\mathscr{C} + c_n(y, u) \le \mathscr{C}_{\max}. \tag{P6.15}$$

Thus introduction of the constraint leads to a recurrence in two variables, which complicates the solution. This equation is to be initialized with $\mathscr{R}_1(y, \mathscr{C})$ given by (P6.10), and defined only on domains $\Omega$ which include $u_1$.

### *P6.5.2 Constraint on Air Consumption*

Let us assume that the movement of water is instantaneous, and that the corresponding air consumption is proportional to

$$\sum_{N}^{n} |F_m - F_{m-1}| \simeq \sum_{N}^{n} |u_m - u_{m-1}| = q_n.$$

Letting $v_n$ be the state variable equal to $u_{n-1}$, and $q_n$ the state variable corresponding to the air consumption, results in the system

$$
\begin{aligned}
&y_{n-1} = e^{-\omega\Delta_n} y_n + (1 - e^{-\omega\Delta_n}) u_n , \\
&v_{n-1} = u_n , \qquad v_N = 0 , \\
&q_{n-1} = q_n + |u_n - v_n| , \qquad q_N = 0 .
\end{aligned}
\tag{P6.16}
$$

The optimal return is now a function of three state variables, and satisfies

$$
\mathscr{R}_n(y, v, q) = \min_{u \in \mathscr{D}} \Big\{ \sum_i H_i \varphi(V_i, y, u) + \mathscr{R}_{n-1}[e^{-\omega\Delta_n} y + (1 - e^{-\omega\Delta_n}) u, u, q + |u - v|] \Big\} . \tag{P6.17}
$$

The domain $\mathscr{D}$ is defined to consist of all $u$ such that

$$
\begin{aligned}
&U_m \leq u \leq U_M , \\
&q + |u - v| \leq Q_{\max} .
\end{aligned}
\tag{P6.18}
$$

The recurrence relation is now in three variables, which still further complicates the solution of (P6.17).

## P6.6 Flow Chart of the Controller

The solution of (P6.8), (P6.14), or (P6.17) yields $u$ as a function of the state variables, and thus results in closed-loop control. In order to detail this result in the form of a flow chart, we shall consider the most complex case, that in which the compressed air consumption is constrained. Thus the control is a function

$$
\hat{u}_n = \mathscr{U}_n[\text{speed, consumption, last control}] , \tag{P6.19}
$$

The index $n$ corresponds to the number of the depth deviation $x_n$ for which $u_n$ is to be calculated by (P6.19).

We assume that there is available a real-time computer with priority interrupt, and that each time $z_f - z \equiv x_n$, an interrupt initiates execution of the program which calculates $u_n$. The corresponding flow chart is shown in Fig. P6.3. The vertical velocity coder can be a pulse counter counting clock pulses, triggered on when $z = z_f - x_n + \Delta$ and off when $z = z_f - x_n$.

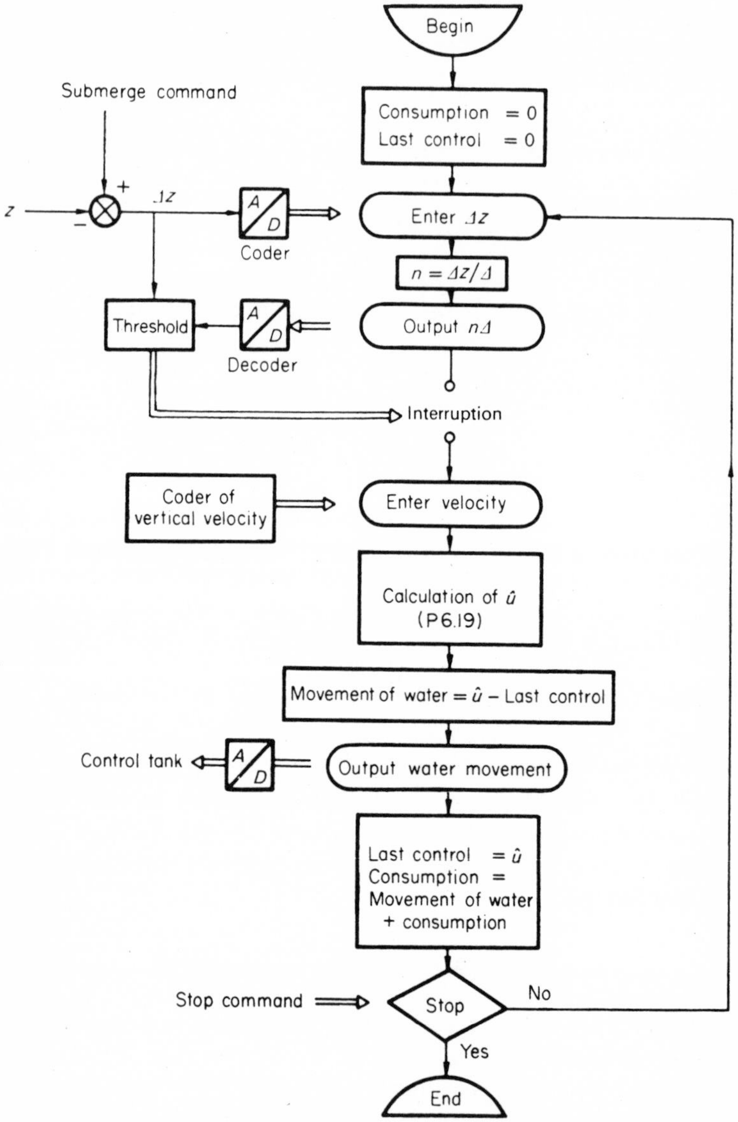


FIG. P6.3. The optimal controller.

# Problem 7 | Optimal Interception

## P7.1 Statement of the Problem

As shown in Fig. P7.1, let us consider an airplane flying in a straight line, with a velocity vector $V$ making an angle $\theta$ with the horizontal. The aircraft exerts a thrust $P(V, H)$ in the direction of the velocity vector, and is subject to a drag $R(V, H)$, with $H$ being the altitude. The mass $m$ of the aircraft varies according to some law

$$dm/dt = -\psi(H, V).$$

Starting from the dynamic equation along $Ox$, and from the kinematic equation in the vertical direction, we shall write the vector equation obeyed by the state vector, with the state variables taken as $V$, $H$, and $m$. The climb angle $\theta$ will be taken to be the control variable. We shall then write this

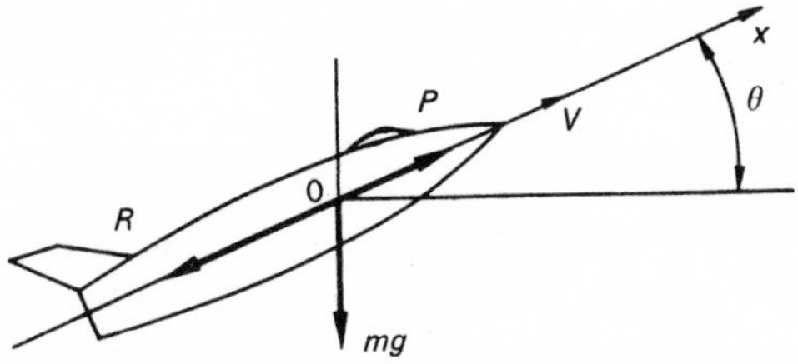


FIG. P7.1. The system to be controlled.

equation in discrete form, assuming that the control interval $\Delta T$ is so small that

$$dx/dt \simeq (x_{n+1} - x_n)/\Delta T.$$

We shall then seek the control policy which carries the aircraft from the state $(V_0, H_0, m_0)$ to the state $(V_f, H_f, m_f)$ in minimum time. This is the interception problem.

After applying the principle of optimality to the discrete model, we shall determine an implicit equation for the optimal policy, and then write a recursive equation which has for its limit the solution of this implicit equation. Then, by letting $\Delta T$ tend to zero, in the discrete implicit equation, we shall establish an implicit partial differential equation from which can be found the form of the optimal law $\hat{u}$ corresponding to the case in which $H(t)$ and $V(t)$ are to be monotonically nondecreasing.

We shall then assume that $m$ is constant, and that only two types of behavior are to be considered: ($\alpha$) $H =$ constant, with a resulting $\Delta V$ in time $\Delta T_\alpha$; ($\beta$) $V =$ constant, with a resulting $\Delta H$ during time $\Delta T_\beta$. (If the $\Delta T$ are small, any control law can be approximated by these two simple laws.) Finally, we shall assume that $H$ and $V$ are quantized, and that only values

$$H_j = H_0 + j\,\Delta H, \qquad 0 \le j \le M,$$

$$V_i = V_0 + i\,\Delta V, \qquad 0 \le i \le N$$

are of interest. A state is then represented by its corresponding pair $(i, j)$, and the controls to be considered are

$$\begin{pmatrix} i \\ j \end{pmatrix} \overset{\alpha}{\to} \begin{pmatrix} i+1 \\ j \end{pmatrix}$$

with a time $T_\alpha\,(i, j)$, and

$$\begin{pmatrix} i \\ j \end{pmatrix} \overset{\beta}{\to} \begin{pmatrix} i \\ j+1 \end{pmatrix}$$

with a time $T_\beta\,(i, j)$. The problem being thus reduced to a problem with discrete states, the principle of optimality may be applied to determine the control to go in minimum time from the initial state $(i, j)$ to the final state $(N, M)$.

## P7.2 Establishment of the Mathematical Model

The equation of motion of the aircraft along the axis $Ox$ is

$$m\dot{V} = P(V, H) - R(V, H) - mg \sin\theta$$

[absorbing a term $Vm$ into $P(V, H)$]. The rate of climb is simply

$$\dot{H} = V \sin\theta,$$

and the equation for the mass loss is, as assumed,

$$\dot{m} = -\psi(H, V).$$

Letting $u = \sin\theta$ be the control, the constraint $0 \leq |u| \leq 1$ is necessary.

Collecting these equations together results in the state equation

$$\begin{aligned} \dot{V} &= [P(V, H) - R(V, H)]/m - gu, \\ \dot{H} &= Vu, \\ \dot{m} &= -\psi(H, V), \end{aligned} \qquad \text{(P7.1)}$$

or, letting $x_1 = V$, $x_2 = H$, $x_3 = m$,

$$\begin{aligned} \dot{x}_1 &= \varphi(x_1, x_2)/x_3 - gu, \\ \dot{x}_2 &= x_1 u, \\ \dot{x}_3 &= -\psi(x_1, x_2), \end{aligned}$$

where

$$\varphi(x_1, x_2) = P(x_1, x_2) - R(x_1, x_2).$$

In discrete form, these equations are

$$\begin{aligned} V^+ &= V + \Delta T[\varphi(V, H)/m - gu], \\ H^+ &= H + \Delta T\, Vu, \\ m^+ &= m - \Delta T\, \psi(V, H). \end{aligned} \qquad \text{(P7.2)}$$

## P7.3 The Equation of Optimality

### *P7.3.1 The Discrete Case*

The problem being stationary, for a given final state the minimum time depends only on the initial state. Letting $\mathcal{T}(H, V, m)$ be the minimum time to go from the state $(H, V, m)$ to the state $(H_f, V_f, m_f)$, the principle of optimality yields

$$\mathscr{T}(H, V, m) = \min_{0 \le |u| \le 1} \{\Delta T + \mathscr{T}(H^+, V^+, m^+)\}, \tag{P7.3}$$

with $\mathscr{T}(H_f, V_f, m_f) = 0$.

Substituting (P7.2) into (P7.3) results in

$$\mathscr{T}(H, V, m) = \Delta T + \min_{0 \le u \le 1} \mathscr{T}[H + \Delta T\, Vu, \\ V + \Delta T(\varphi(V, H)/m - gu), m - \Delta T\, \psi(V, H)], \tag{P7.4}$$

which is an implicit equation for the unknown function $\mathscr{T}(H, V, m)$. This function is then the limit of the sequence generated by the recursion

$$\mathscr{T}_{n+1}(H, V, m) = \Delta T + \min_{0 \le u \le 1} \mathscr{T}_n[H + \Delta T\, Vu, V + \Delta T(\varphi(V, H)/m - gu), \\ m - \Delta T\, \psi(V, H)], \qquad \mathscr{T}_0(H_f, V_f, m_f) = 0. \tag{P7.5}$$

### *P7.3.2 Passage to the Continuous Case*

If $\Delta T$ is small, the implicit relation (P7.3) can be written

$$\mathscr{T}(H, V, m) = \Delta T + \min_u [\mathscr{T}(H, V, m) + (\partial \mathscr{T}/\partial H)\dot{H}\,\Delta T + (\partial \mathscr{T}/\partial V)\,\dot{V}\Delta T \\ + (\partial \mathscr{T}/\partial m)\dot{m}\,\Delta T] + O(\Delta^2).$$

Using the expressions for $V$, $H$, and $m$, this becomes

$$\mathscr{T} - \Delta T = \min_u [\mathscr{T} + (\partial \mathscr{T}/\partial H)Vu\,\Delta T + (\partial \mathscr{T}/\partial V)(\varphi/m - gu)\Delta T \\ + (\partial \mathscr{T}/\partial m)(-\psi)\,\Delta T],$$

or

$$0 = \Delta T[1 + (\partial \mathscr{T}/\partial V)(\varphi/m) - (\partial \mathscr{T}/\partial m)\psi] \\ + \Delta T \min_u [(\partial \mathscr{T}/\partial H)V - (\partial \mathscr{T}/\partial V)g]u.$$

This yields finally the implicit equation

$$1 + \frac{\partial \mathscr{T}(H, V, m)}{\partial V}\frac{\varphi(H, V)}{m} - \frac{\partial \mathscr{T}(H, V, m)}{\partial m}\psi(H, V) \\ = \min_{0 \le |u| \le 1} \left[\frac{\partial \mathscr{T}(H, V, m)}{\partial V} g - \frac{\partial \mathscr{T}(H, V, m)}{\partial H} V\right] u. \tag{P7.6}$$

The solution of this last equation is impossible to find in literal form, and only the recurrence equation found in the discrete case leads to a numerical solution.

### *P7.3.3 Form of the Optimal Control Law*

Even though the partial differential equation (P7.6) does not lead to a detailed solution, it does allow the form of $\hat{u}$ to be found. The minimum of the indicated quantity occurs for

$$\hat{u} = -1 \quad \text{if} \quad (\partial\mathcal{T}/\partial V)g - (\partial\mathcal{T}/\partial H)V > 0 \Rightarrow \hat{\theta} = -\pi/2, \tag{P7.7}$$

$$\hat{u} = 1 \quad \text{if} \quad (\partial\mathcal{T}/\partial V)g - (\partial\mathcal{T}/\partial H)V < 0 \Rightarrow \hat{\theta} = \pi/2. \tag{P7.8}$$

In the particular case that $V$ and $H$ are to be nondecreasing, $\theta$ is bounded by $\theta_{\min} = 0$, and a value $\theta_{\max}$ determined by $\dot{V} = 0$. The optimal policy then consists of stretches of level flight to gain speed, alternated with constant-speed climbs.

## P7.4 Discrete State Space

We shall show that discretization of the state space leads to a simple solution for the problem. The control law which results is in good agreement with that described in the preceding paragraph. In addition, this law is such that the grid in state space applies whatever the combination of controls used.

### *P7.4.1 Calculation of the Transition Times*

The first type of control law, such that $H =$ constant, leads to

$$\dot{H} = 0 \rightarrow Vu = 0 \Rightarrow u = 0,$$

$$\Delta V/\Delta T \simeq \dot{V} = \varphi(H, V)/m, \tag{P7.9}$$

$$\Delta T_\alpha(H, V) = \Delta V m/\varphi(H, V).$$

The second law, $V =$ constant, leads to

$$\dot{V} = 0 \rightarrow \varphi/m - gu = 0 \Rightarrow u = \varphi/mg.$$

It is necessary to check here that $|u| \leq 1$. Then

$$\begin{aligned} \Delta H/\Delta T \simeq \dot{H} &= Vu = V\varphi/mg, \\ \Delta T_\beta(H, V) &= mg[\Delta H/V\varphi(H, V)]. \end{aligned} \tag{P7.10}$$

These relations are summarized in Fig. P7.2.

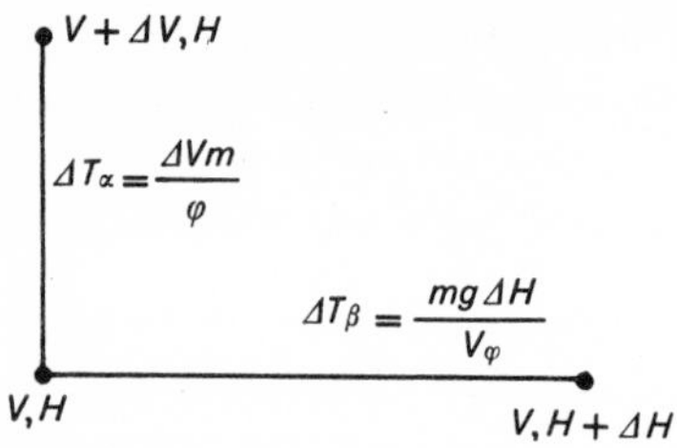


FIG. P7.2. System transition times.

Using (P7.9) and (P7.10), we then have

$$\begin{aligned} T_\alpha(i,j) &= \Delta V\, m/\varphi(H_0 + j\,\Delta H,\ V_0 + i\,\Delta V), \\ T_\beta(i,j) &= \Delta H\, mg/[(V_0 + i\ \Delta V)\varphi(H_0 + j\,\Delta H,\ V_0 + i\,\Delta V)]. \end{aligned} \tag{P7.11}$$

### *P7.4.2 The Equation for Optimality*

Letting $\mathscr{T}_{ij}$ be the optimal time starting from the initial state $V_i$, $H_j$, the principle of optimality leads to

$$\begin{aligned} \mathscr{T}_{ij} &= \min_{(\alpha,\beta)}[T_\alpha(i,j) + \mathscr{T}_{i+1,j} \mid T_\beta(i,j) + \mathscr{T}_{i,j+1}], \\ \mathscr{T}_{NM} &= 0. \end{aligned} \tag{P7.12}$$

Apparently we again have to deal with an implicit equation. Actually, however, the values $\mathscr{T}_{ij}$, for $i + j = k$, can be calculated from the values $\mathscr{T}_{i'j'}$ for $i' + j' = k + 1$. Thus, since $\mathscr{T}_{NM} = 0$, this equation can be solved directly for $k = N + M, N + M - 1, \ldots$. In fact, this property shows that the control law obtained only allows increases in $i$ or $j$.

If climb with loss of speed or altitude is permitted, this equation becomes implicit and can only be solved by iteration. Note that the solution obtained here associates with each pair $(i, j)$ the optimal control, and thus leads to the optimal interception policy. The solution of (P7.12) is illustrated for one case in Fig. P7.3.

### *P7.4.3 An Optimal Intercept Calculator*

In Fig. P7.4 there is shown the arrangement of an optimal intercept calculator, using the above results. The problem is to control an interceptor in such a way that, in the shortest possible time, the aircraft is in a favorable position to attack some target.

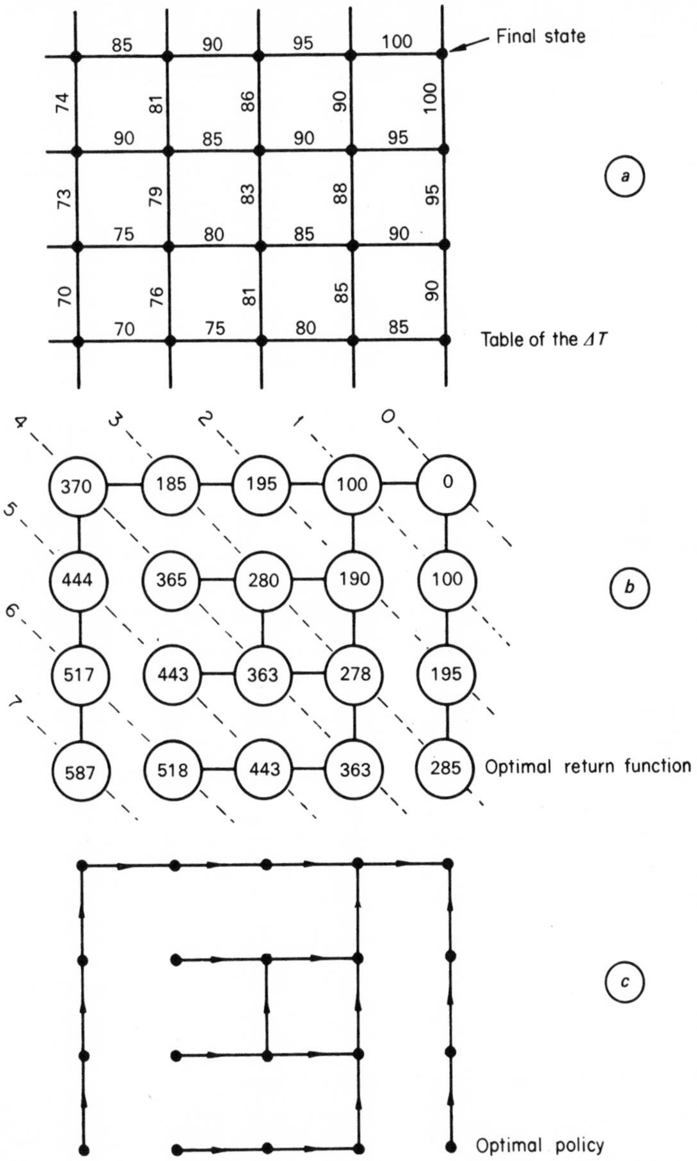


FIG. P7.3. An example of solution of the optimality equation.

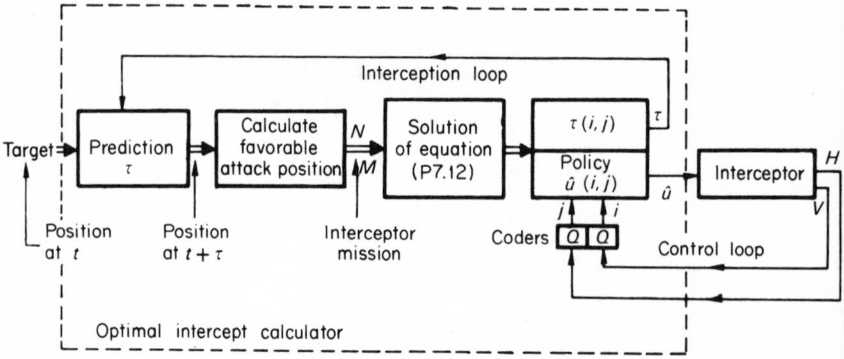


FIG. P7.4. The optimal intercept calculator.

Given information about the most recent position of the target, the target position is extrapolated to time $t + \tau$, which determines the position $(N, M)$ which the interceptor should occupy at that time. (This calculation involves a secondary optimization problem.) Solution of the recurrence equation (P7.12) then furnishes, for this $(N, M)$, the minimum time $\tau$ at the end of which the interceptor will be in this final position. The final attack position is improved further, iteratively, using an interception loop.

The control policy resulting from this loop is used in the interceptor controller along with quantized values of the altitude and speed of the interceptor.

# Problem 8 | Control of a Continuous Process

## P8.1 Statement of the Problem

Let us consider a body in motion, with displacement $s$ described by

$$\ddot{s} = u,$$

where the control force $u$ is such that $|u| \leq U$. We wish to find the optimal control law, which carries the body to the final position $s(T) = d$ at a given time $T$, in such a way that $\dot{s}(T)$ is maximum.

To solve this problem, we first establish the corresponding optimality equation, and determine from it the general form of the optimal control. Direct solution of the optimality equation found is difficult, and for that reason we next study the simplest law of the type indicated (a "bang-bang" law), and show that the corresponding return function satisfies the optimality equation. We then apply the principle of optimality to obtain directly the exact form of the optimal control law. Finally, the transversality conditions supply the sign of the first control which is to be applied.

To complete the study, we examine the practical synthesis of this optimal controller. We use the properties of the adjoint system, considering first the case of an arbitrary process equation, and then the special case of

$$\ddot{s} + a\dot{s} = u.$$

## P8.2 The Optimality Equation

The process

$$\ddot{s} = u$$

corresponds to the state-vector model

$$\dot{\boldsymbol{x}} = \begin{bmatrix} 0 & 1 \\ 0 & 0 \end{bmatrix} \boldsymbol{x} + \begin{bmatrix} 0 \\ 1 \end{bmatrix} u, \qquad s = [1 \quad 0]\boldsymbol{x}, \tag{P8.1}$$

where

$$x_1 = s, \qquad x_2 = \dot{s}.$$

In this problem, only a terminal cost is present. Thus no elementary return appears in the optimality equation, and the optimal return function is simply the cost assigned to the final domain at the terminal instant.

Let $\mathscr{R}(\boldsymbol{x}, \theta)$ be the optimal return (final velocity) corresponding to an initial state $\boldsymbol{x}$ and a time-to-go of $\theta = T - t$ sec. Then

$$\hat{\mathscr{R}}(\boldsymbol{x}, \theta) = \max_u [\hat{\mathscr{R}}(\boldsymbol{x} + \Delta\boldsymbol{x}, \theta - \varDelta)], \tag{P8.2}$$

where $\boldsymbol{x} + \Delta\boldsymbol{x}$ is the state attained after $\varDelta$ seconds, starting from state $\boldsymbol{x}$ and applying the control $u$. If $\varDelta$ is small, the term in brackets can be written

$$[\,] \simeq \hat{\mathscr{R}}(\boldsymbol{x}, \theta) + \hat{\mathscr{R}}_x{}^{\mathrm{T}} \dot{\boldsymbol{x}} \varDelta - \hat{\mathscr{R}}_\theta \varDelta .$$

After passing to the limit and simplifying, there results

$$\partial \hat{\mathscr{R}}(\boldsymbol{x}, \theta)/\partial\theta = \max_u [\hat{\mathscr{R}}_x{}^{\mathrm{T}} \dot{\boldsymbol{x}}].$$

For our problem, this becomes

$$\partial\hat{\mathscr{R}}/\partial\theta = \max_u [(\partial\hat{\mathscr{R}}/\partial x_1)x_2 + (\partial\hat{\mathscr{R}}/\partial x_2)u], \tag{P8.3}$$

with the terminal condition

$$\hat{\mathscr{R}}(d, v, 0) \equiv v, \qquad \forall v . \tag{P8.4}$$

## P8.3 Search for the Optimal Policy

The optimality equation (P8.3) is linear in $u$. Thus, since $|u| \le U$, the optimal control law is

$$\hat{u} = U \operatorname{sign}[\partial\hat{\mathscr{R}}/\partial x_2] \rightarrow \hat{u} = \pm U .$$

This result, in fact, holds true for any linear process.

Using this optimal control in (P8.3), the optimality equation becomes

$$\partial\hat{\mathscr{R}}/\partial\theta = (\partial\hat{\mathscr{R}}/\partial x_1)x_2 + |\partial\hat{\mathscr{R}}/\partial x_2|\,U, \qquad \hat{\mathscr{R}}(d, v, 0) \equiv v, \quad \forall v. \quad \text{(P8.5)}$$

This is a nonlinear partial differential equation, and it is not known how to calculate its solution directly in literal form. However, since the general form of the solution is known, we can assume some particular control law, calculate the corresponding return, and check whether the boundary conditions and the optimality equation are satisfied for that control.

## P8.4 Study of a Bang-Bang Law

Let us study the simplest possible law,

$$\hat{u}(t) = \varepsilon, \qquad 0 \le t \le \tau$$

$$\hat{u}(t) = -\varepsilon, \qquad \tau < t \le \theta$$

with $\varepsilon = \pm U$. To calculate the terminal state and apply the condition that the terminal condition $s = d$ be satisfied, we can integrate the state equation with this control as input. For $u = \varepsilon$, the result of this integration is

$$\boldsymbol{x}(t) = \begin{bmatrix} 1 & t \\ 0 & 1 \end{bmatrix} \boldsymbol{x}(0) + \begin{bmatrix} \dfrac{t^2}{2} \\ t \end{bmatrix} \varepsilon. \qquad \text{(P8.6)}$$

This leads to the final solution, taking account of the switching of the control at time $\tau$,

$$\boldsymbol{x}(\theta) = \begin{bmatrix} 1 & (\theta - \tau) \\ 0 & 1 \end{bmatrix} \left\{ \begin{bmatrix} 1 & \tau \\ 0 & 1 \end{bmatrix} \boldsymbol{x}(0) + \begin{bmatrix} \dfrac{\tau^2}{2} \\ \tau \end{bmatrix} \varepsilon \right\} - \begin{bmatrix} \dfrac{(\theta - \tau)^2}{2} \\ \theta - \tau \end{bmatrix} \varepsilon$$

$$= \begin{bmatrix} 1 & \theta \\ 0 & 1 \end{bmatrix} \boldsymbol{x}(0) + \begin{bmatrix} \dfrac{\theta^2}{2} - (\theta - \tau)^2 \\ 2\tau - \theta \end{bmatrix} \varepsilon. \qquad \text{(P8.7)}$$

Applying the terminal constraint to the motion described by (P8.5) results in the requirement

$$d = x_1 + \theta x_2 + \theta^2\varepsilon/2 - (\theta - \tau)^2\varepsilon, \qquad \text{(P8.8)}$$

so that the switching time must satisfy

$$\tau = \theta - \{[(x_1 + \theta x_2 - d)/\varepsilon] + \theta^2/2\}^{1/2}. \qquad \text{(P8.9)}$$

Since $0 < \tau \leq \theta$, in order that this have a solution it is necessary that

$$0 \leq [(x_1 + \theta x_2 - d)/\varepsilon] + \theta^2/2 \leq \theta^2,$$

$$|x_1 + \theta x_2 - d| \leq \varepsilon\theta^2/2. \qquad \text{(P8.10)}$$

Finally, the return corresponding to this control law can be written

$$\mathscr{R}(\mathbf{x}, \theta) \equiv x_2(\theta) = x_2 + \theta\varepsilon - 2\varepsilon\{[(x_1 + \theta x_2 - d)/\varepsilon] + \theta^2/2\}^{1/2}. \quad \text{(P8.11)}$$

We now wish to check whether there exists a value of $\varepsilon(= \pm U)$ for which the solution (P8.7) satisfies the partial differential equation (P8.5). The boundary condition (P8.4) is satisfied, since from (P8.11) it is clear that

$$\mathscr{R}(d, v, 0) \equiv v.$$

Proceeding from (P8.11), we have

$$\partial\mathscr{R}/\partial\theta = \varepsilon - \varepsilon\{[(x_2/\varepsilon) + \theta]/\sqrt{\ \ }\} = \varepsilon - [(x_2 + \varepsilon\theta)/\sqrt{\ \ }],$$

$$\partial\mathscr{R}/\partial x_1 = -1/\sqrt{\ \ },$$

$$\partial\mathscr{R}/\partial x_2 = 1 - \theta/\sqrt{\ \ },$$

where the radical symbol indicates the square root factor in (P8.11), from which

$$\varepsilon - [(x_2 + \varepsilon\theta)/\sqrt{\ \ }] \stackrel{?}{=} -(x_2/\sqrt{\ \ }) + |1 - (\theta/\sqrt{\ \ })|\, U,$$

$$\varepsilon[1 - (\theta/\sqrt{\ \ })] \stackrel{?}{=} |1 - (\theta/\sqrt{\ \ })|\, U.$$

This equality will hold true if

$$\varepsilon = U \operatorname{sign}(1 - \theta/\sqrt{\ \ }).$$

But $\tau \geq 0 \to \theta \geq \sqrt{\ \ }$, so that we should choose $\hat{\varepsilon} = -U$. Then

$$\hat{\tau} = \theta - \{[(-x_1 - \theta x_2 + d)/U] + (\theta^2/2)\}^{1/2}. \qquad \text{(P8.12)}$$

Switching of the control should take place when $\hat{\tau} = 0$, which implies that $d$ is just attainable using the control $+U$. Then

$$\hat{\tau} = 0 \Rightarrow U\theta^2/2 = d - (x_1 + \theta x_2). \qquad \text{(P8.13)}$$

The final results of this study are that the optimal law is

$$\hat{u} = U \operatorname{sign}[d - U\theta^2/2 - (x_1 + x_2\theta)], \qquad \text{(P8.14)}$$

and the controllable region is

$$|x_1 + x_2\theta - d| \leq U\theta^2/2.$$

## P8.5 Use of the Maximum Principle

The results used in this paragraph are discussed further by Boudarel *et al.* [1, Vol. 4].

The adjoint system for this problem is

$$\dot{\psi} = -\begin{bmatrix} 0 & 0 \\ 0 & 1 \end{bmatrix} \psi, \tag{P8.15}$$

which, when integrated, yields

$$\psi(t) = \begin{bmatrix} \alpha \\ \beta - \alpha t \end{bmatrix}. \tag{P8.16}$$

The constants of integration, $\alpha$ and $\beta$, should be such that the terminal conditions on $\psi(\theta)$ are satisfied. Since in this problem there is a terminal constraint on the state, the constraint on $\psi(\theta)$ is of the form

$$\psi(\theta) = W_1 + W_2, \tag{P8.17}$$

where

$$W_1 = \begin{bmatrix} 0 \\ 1 \end{bmatrix}$$

is a vector in the direction of the terminal constraint, and

$$W_2 = \begin{bmatrix} k \\ 0 \end{bmatrix}$$

is a vector normal to the terminal constraint space. The vector $\psi(\theta) - W_1$ can be interpreted as the adjoint vector obtained with an elementary return which is zero for $t < \theta$, and an impulse of area $x_2$ at the instant $\theta$. The vector $W_2$ is the classical transversality condition. Applying the condition (P8.17) results in

$$\psi_2(\theta) = \beta - \alpha\theta = 1 .$$

The Hamiltonian for this problem is

$$H = \psi_1 \dot{x}_1 + \psi_2 \dot{x}_2 = \psi_1 x_2 + \psi_2 u . \tag{P8.18}$$

Its maximum is attained for

$$\hat{u} = +U \operatorname{sign}(\psi_2). \tag{P8.19}$$

But the function $\psi_2(t)$ is linear, and thus has only one zero. In addition, $\psi_2(\theta) = +1$. Thus the optimal control law can only be

$$\begin{aligned} 0 \le t \le \tau, \quad & \hat{u} = -U, \\ \tau \le t \le T, \quad & \hat{u} = +U, \end{aligned} \tag{P8.20}$$

in the case that a change of sign occurs on the interval $[0, T]$.

Thus the maximum principle leads directly to the bang-bang solution, starting with $u = -U$. The subsequent calculation of the switching condition is, however, identical to that carried out in the preceding section.

## P8.6 Synthesis of the Controller

We have seen that the optimal policy consists of applying $u = +U$, beginning at some time $\tau$ determined by the state, and that this policy results in $s(\theta) = d$. The determination of the switching time $\tau$, carried out in Section P8.4 using a literal method, can be carried through analogously for any second-order transfer function.

To see this, consider an arbitrary system of the form

$$\begin{aligned} \dot{\boldsymbol{x}} &= \mathbf{A}\boldsymbol{x} + \boldsymbol{B}u, \\ s &= \boldsymbol{c}^{\mathrm{T}}\boldsymbol{x}. \end{aligned} \tag{P8.21}$$

The solution is

$$\boldsymbol{x}(T) = \boldsymbol{\varphi}(T - t)\boldsymbol{x}(t) + U\int_0^{T-t} \boldsymbol{\varphi}(\alpha)\boldsymbol{B}\, \mathrm{d}\alpha, \tag{P8.22}$$

where $\boldsymbol{\varphi}(\alpha)$ satisfies

$$\mathrm{d}\boldsymbol{\varphi}(\alpha)/\mathrm{d}\alpha = \mathbf{A}\boldsymbol{\varphi}(\alpha), \tag{P8.23}$$

$$\boldsymbol{\varphi}(0) = \mathbf{1}. \tag{P8.24}$$

Thus

$$s(T) = \boldsymbol{c}^{\mathrm{T}}\boldsymbol{\varphi}(T - t)\boldsymbol{x}(t) + U\int_0^{T-t} \boldsymbol{c}^{\mathrm{T}}\boldsymbol{\varphi}(\alpha)\boldsymbol{B}\, \mathrm{d}\alpha. \tag{P8.25}$$

The final time is $T$ in absolute time, or $\theta = T - t = 0$ in relative time. Let

$$\boldsymbol{v}(t) = \boldsymbol{\varphi}^{\mathrm{T}}(T - t)\boldsymbol{c} = \boldsymbol{\varphi}^{\mathrm{T}}(\theta)\boldsymbol{c}, \tag{P8.26}$$

$$K(t) = \int_0^{T-t} \boldsymbol{c}^{\mathrm{T}}\boldsymbol{\varphi}(\alpha)\boldsymbol{B}\, \mathrm{d}\alpha. \tag{P8.27}$$

From (P8.25), the output at time $T$, with an initial state $\boldsymbol{x}(t)$, is

$$s(T) \equiv \boldsymbol{v}^{\mathrm{T}}(t)\boldsymbol{x}(t) + UK(t). \tag{P8.28}$$

Using (P8.23), and (P8.25), these quantities can be shown to satisfy the adjoint linear differential equations

$$\mathrm{d}\boldsymbol{v}(t)/\mathrm{d}t = -\mathbf{A}^{\mathrm{T}}\boldsymbol{v}(t), \qquad \boldsymbol{v}(T) = \boldsymbol{c}, \tag{P8.29}$$

$$\mathrm{d}K(t)/\mathrm{d}t = -\boldsymbol{v}^{\mathrm{T}}(t)\boldsymbol{B}, \qquad K(T) = 0. \tag{P8.30}$$

The two terminal conditions can be written

$$s(T) = \boldsymbol{c}^{\mathrm{T}}\boldsymbol{x}(T).$$

The initial condition $\boldsymbol{v}(0)$, such that the terminal condition $\boldsymbol{v}(T) = \boldsymbol{c}$ is satisfied, can be obtained by backward integration of the differential system for $\boldsymbol{v}$. The same is true for the condition on $K(0)$.

The scheme shown in Fig. P8.1 summarizes these calculations, and indicates the form of the optimal controller.

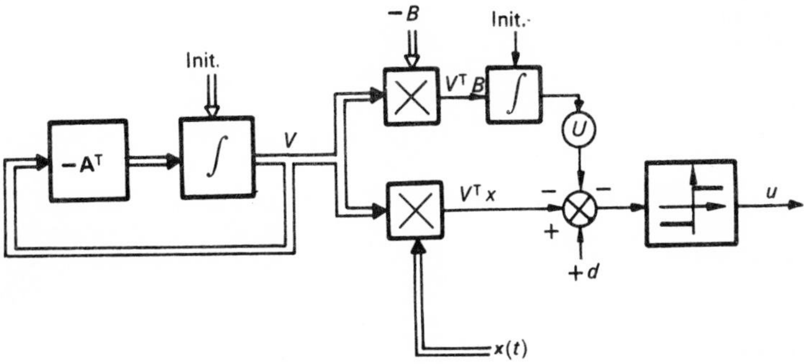


FIG. P8.1. The optimal controller of a continuous system.

In the special case

$$\ddot{s} + a\dot{s} = u, \tag{P8.31}$$

$$\mathbf{A} = \begin{bmatrix} 0 & 1 \\ 0 & -a \end{bmatrix}, \qquad \boldsymbol{B} = \begin{bmatrix} 0 \\ 1 \end{bmatrix}, \qquad \boldsymbol{c} = \begin{bmatrix} 1 \\ 0 \end{bmatrix},$$

so that $\boldsymbol{v}^{\mathrm{T}}\boldsymbol{B} = v_2(t)$ and

$$\dot{v}_1(t) = 0 \rightarrow v_1(t) = \alpha,$$

$$\dot{v}_2(t) = -\alpha + av_2(t) \rightarrow v_2(t) = (\alpha/a) + \beta e^{at}.$$

The terminal condition can be expressed algebraically as

$$\begin{aligned} \alpha &= 1 \\ \alpha/a + \beta e^{aT} &= 0 \end{aligned} \quad \beta = -e^{-aT}/a,$$

from which

$$v(0) = \begin{bmatrix} 1 \\ \dfrac{1 - e^{-aT}}{a} \end{bmatrix}.$$

In the same way,

$$\dot{K}(t) = -v_2(t) = -(1 - e^{a(t-T)})/a \rightarrow K(t) = \gamma - (t/a) + (e^{a(t-T)}/a^2),$$

$$K(T) = 0 \rightarrow \gamma = (T/a) - (1/a^2),$$

from which

$$K(0) = (-1/a^2) + (T/a) + (e^{-aT}/a^2).$$

# Appendix | Filtering

In Chapter 9, we saw that synthesis of the optimal controller required formation of an estimate $\hat{\boldsymbol{x}}_n$ of the state vector. This calculation belongs to the general problem of filtering and prediction, the principal results of which we summarize here.

We first define the filtering problem, and after introducing the concept of an optimal estimate, state some classical theorems from the theory of random functions. These in turn lead to the principal methods for solution of the problem.

In this appendix we confine ourselves entirely to the methods developed by Kalman. The reader may consult Boudarel *et al.* [1, Vol. 4], for a detailed account of the Wiener–Newton methods.

Having in mind the derivation of the discrete Kalman filter, we give first some brief discussion of the geometrical representation of random variables, which permits reducing the filtering problem to a geometrical projection problem. We stress the recursive nature of the Kalman filter, and its implementation. In closing, we briefly indicate the extension of the discrete results to the continuous case.

## A.1 The Filtering and Prediction Problems

### *A.1.1 Definition of the Problem*

In Fig. A.1 there is a schematic representation of the problem in the continuous case. The useful signal is $\boldsymbol{e}(t)$, and $\boldsymbol{b}(t)$ is a perturbation. The

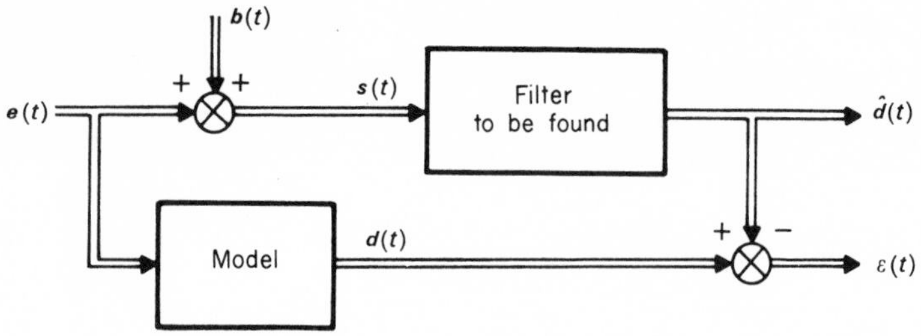


FIG. A.1. The filtering problem.

signal to be filtered to form an estimate $\hat{d}(t)$ is $s(t) = e(t) + b(t)$. The exact value is $d(t)$, the quantity to be reconstructed as well as possible by the filter. It depends deterministically on the signal through some functional indicated as the model.

Depending on the nature of the functional used as the model, various particular problems can be distinguished. If $d(t) = e(t)$, the problem is one of estimation. If $d(t) = e(t + \tau)$, prediction is required. Estimation of the derivative of the signal is indicated by $d(t) = \dot{e}(t)$, power measurement by $d(t) = [e(t)]^2$, energy measurement by $d(t) = \int_0^T e^2(t)\,dt$, and so forth.

All these various signals are defined by means of statistical models, either correlation functions or Markov process generators.

### *A.1.2 Definition of an Optimum Filter*

In order to characterize the quality of the filter operation at the instant $t$, the error $\varepsilon(t) = d(t) - \hat{d}(t)$ is introduced, as well as a return function $L(\varepsilon)$. Since the signals involved are random, $L(\varepsilon)$ is itself a random variable, and can not be used directly to compare filters. The quality criterion used is thus taken as the mathematical expectation of $L(\varepsilon)$. Since the knowledge of $s(\tau)$ for $\tau \leq t$ is clearly of importance, since it restricts the domain of possible values $d(t)$, these values should be taken account of in the filtering, and the mathematical expectation used in defining the quality criterion should be conditional on $s(\tau)$ for $\tau \leq t$.

This formulation of the problem is the classical one, assuming that all statistical quantities in the mathematical models of $e(t)$ and $b(t)$ are completely known. In fact, in some problems, certain of these parameters may not be known, and the above criterion has no meaning. The most natural approach then is to seek the least bad solution, which leads to a minimax problem. We shall not pursue these questions further, but only refer the reader to more specialized works.

In the above discussion, we have not been precise about the nature of the filter sought. It is clear that $\hat{\boldsymbol{d}}(t)$ should depend on the values $\boldsymbol{s}(\tau)$ for $\tau \leq t$, but this dependence can be either deterministic, in which case $\hat{\boldsymbol{d}}(t)$ is a functional of $\boldsymbol{s}(\tau)$ for $\tau \leq t$, or stochastic, in which case the probability law defining $\hat{\boldsymbol{d}}(t)$ depends on $\boldsymbol{s}(\tau)$. However, in the majority of cases and in particular when $L(\boldsymbol{\varepsilon})$ is convex, the optimal stochastic law reduces to a deterministic dependence. In the following, we shall restrict ourselves to that case.

### *A.1.3 Review of Theoretical Results*

We shall state three theorems from the theory of random functions [28].

**Theorem 1.** *If the conditional probability distribution*

$$P(\boldsymbol{x}) = \text{prob}\{\boldsymbol{d}(t) \leq \boldsymbol{x} \,|\, \boldsymbol{s}(\tau) \quad \text{for} \quad \tau \leq t\},$$

*is symmetric around the mean* $\hat{\boldsymbol{d}}$ *and convex in the region* $\boldsymbol{x} \leq \hat{\boldsymbol{d}}$, *then the best estimate* $\hat{\boldsymbol{d}}(t)$ *is the conditional mathematical expectation of* $\boldsymbol{d}(t)$. *This result is independent of* $L(\boldsymbol{\varepsilon})$, *except that it must satisfy the conditions:*

$$L(\boldsymbol{0}) = 0,$$

$$L(\boldsymbol{\varepsilon}) = L(-\boldsymbol{\varepsilon}) \geq 0,$$

$$\|\boldsymbol{\varepsilon}_1\| \leq \|\boldsymbol{\varepsilon}_2\| \Rightarrow L(\boldsymbol{\varepsilon}_1) \leq L(\boldsymbol{\varepsilon}_2).$$

**Theorem 2.** *If* $L(\boldsymbol{\varepsilon}) = \|\boldsymbol{\varepsilon}\|^2$, *where* $\| \ \|^2$ *indicates a quadratic norm, then, whatever the function* $P(\boldsymbol{x})$ *may be, the optimal estimate* $\hat{\boldsymbol{d}}(t)$ *is the conditional mathematical expectation of* $\boldsymbol{d}(t)$.

**Theorem 3.** *If* $\boldsymbol{d}(t)$ *and* $\boldsymbol{s}(t)$ *are Gaussian, with zero mean, then the best estimate* $\hat{\boldsymbol{d}}(t)$ *is a linear functional of* $\boldsymbol{s}(\tau)$.

The practical consequences of these theorems are great. Theorem 3, for example, allows the field of search to be restricted to linear filters, which justifies the classical approach of Wiener and Newton. As far as the estimation problem posed in Chapter 10 is concerned, if the initial state and the perturbations acting on the process are Gaussian, then all signals in the system will also be Gaussian, due to the linearity of the system. In this case, Theorem 1 indicates that this estimation problem is equivalent to the problem of optimal estimation minimizing a quadratic criterion, which has a linear solution according to Theorems 2 and 3. We shall see that this minimization can be interpreted geometrically as a projection. This approach is more intuitive than that in which the conditional probability law is calculated and used to determine the conditional expectation.

### *A.1.4 Geometrical Representation of Random Variables*

Let $x_i$ be a set of zero-mean random variables, to each of which there corresponds an element $\mathsf{X}_i$ of an abstract space $\mathscr{E}$. Since every linear combination

$$y = \sum_{i=1}^{k} a_i x_i$$

defines a new random variable, this space is a vector space. As the $a_i$ in this combination vary, a vector subspace of $k$ dimensions is generated.

Let us now define a scalar product on this vector space by

$$(\mathsf{X}, \mathsf{Y}) = \mathrm{E}(xy),$$

which leads to the norm

$$\|\mathsf{X}\|^2 = \mathrm{E}(x^2).$$

If attention is now restricted to random variables with finite variance, the abstract space $\mathscr{E}$ is a complete normed vector space, which is to say that it is a Hilbert space.

If an arbitrary vector $\mathsf{V}$ and the subspace $\mathscr{X}$ generated by the variables $x_i$ are now considered, it can be shown that, just as in the case of a Euclidean vector space, the vector $\mathsf{V}$ can be decomposed uniquely into a vector $\hat{\mathsf{V}}$ in $\mathscr{X}$, and a vector $\tilde{\mathsf{V}}$ orthogonal to every vector in $\mathscr{X}$:

$$\mathsf{V} = \hat{\mathsf{V}} + \tilde{\mathsf{V}} \Rightarrow v = \hat{v} + \tilde{v}.$$

The random variable $\hat{v}$ is called the orthogonal projection of $v$ on the subspace generated by the variables $x_i$. We then have:

**Theorem 4.** *The projection $\hat{v}$ is the linear combination*

$$y = \sum_{i=1}^{k} a_i x_i,$$

*which minimizes* $\mathrm{E}[(v - y)^2]$.

To verify this result, let $y = \hat{v} + w$, where $w$ corresponds to some vector $\mathsf{W}$ in $\mathscr{X}$. Then

$$\mathrm{E}[(v - y)^2] = \mathrm{E}(v - \hat{v} - w)^2] = \mathrm{E}[\tilde{v}^2] + \mathrm{E}[w^2] + 2\mathrm{E}[\tilde{v}w].$$

But

$$\mathrm{E}[\tilde{v}w] = 0,$$

since $\mathsf{W} \in \mathscr{X}$ and $\tilde{\mathsf{V}} \perp \mathscr{X}$. Then

$$\mathrm{E}[(v - y)^2] = \mathrm{E}[\tilde{v}^2] + \mathrm{E}[w^2] \geq \mathrm{E}[\tilde{v}^2].$$

The minimum is thus attained for $w = 0$, as claimed.

The main reason for interest in the abstract space just introduced is that it allows the correlations between random variables to be interpreted geometrically.

The formalism which we have introduced establishes a correspondence between a scalar $x$ and a vector $\mathsf{X}$ of the abstract space. In the following, we shall need to consider random vectors $\boldsymbol{x} \in \mathbf{R}^m$. Thus, to each component $x_i$ of $\boldsymbol{x}$ there corresponds a vector $\mathsf{X}_i \in \mathscr{E}$, and thus to $\boldsymbol{x}$ there correspond $m$ vectors of $\mathscr{E}$, constituting a tensor $\mathsf{X}$. The scalar products of these vectors then correspond to the elements of the covariance matrix of $\boldsymbol{x}$:

$$\varphi_{ij} = \mathrm{E}[x_i x_j] = (\mathsf{X}_i, \mathsf{X}_j).$$

That a random vector $\boldsymbol{x}$ is uncorrelated with $\boldsymbol{y}$ means

$$\mathrm{E}[x_i y_j] = 0, \qquad \forall i, j,$$

which is to say that all the $\mathsf{X}_i$ are orthogonal to all the $\mathsf{Y}_j$, or, geometrically, that the projection of $\boldsymbol{y}$ on the subspace generated by $\boldsymbol{x}$ (spanned by the $x_i$) is zero.

In practice, no formal distinction is made between $x_i$ and $\mathsf{X}_i$, and rather than $\mathsf{X}_i \in \mathscr{E}$ we shall write simply $\boldsymbol{x} \in \mathscr{E}$ in the following.

## A.2 Estimation of the State of a Discrete Process

### *A.2.1 Definition of the Problem*

Let us consider the discrete linear process

$$\begin{aligned} \boldsymbol{x}_{n+1} &= \mathbf{F}_n \boldsymbol{x}_n + \mathbf{H}_n \boldsymbol{u}_n + \boldsymbol{\delta}_n, \\ \boldsymbol{s}_n &= \mathbf{G}_n \boldsymbol{x}_n + \boldsymbol{\delta}_n', \end{aligned} \tag{A.1}$$

where $\boldsymbol{x}_n$ is the state, $\boldsymbol{s}_n$ the measurable output, $\boldsymbol{u}_n$ the control, and $\boldsymbol{\delta}_n$ and $\boldsymbol{\delta}_n'$ uncorrelated Gaussian random vectors, such that

$$\mathrm{E}[\boldsymbol{\delta}_n \boldsymbol{\delta}_m^{\mathrm{T}}] = \mathbf{0}, \qquad n \neq m, \qquad \mathrm{E}[\boldsymbol{\delta}_n \boldsymbol{\delta}_n^{\mathrm{T}}] = \boldsymbol{\Phi}_n, \tag{A.2}$$

$$\mathrm{E}[\boldsymbol{\delta}_n' \boldsymbol{\delta}_m'^{\mathrm{T}}] = \mathbf{0}, \qquad n \neq m, \qquad \mathrm{E}[\boldsymbol{\delta}_n' \boldsymbol{\delta}_n'^{\mathrm{T}}] = \boldsymbol{\Phi}_n', \tag{A.3}$$

$$\mathrm{E}[\boldsymbol{\delta}_n \boldsymbol{\delta}_m'^{\mathrm{T}}] = \mathbf{0}, \qquad \forall n, m. \tag{A.4}$$

At time $n$ the available information is

$$\boldsymbol{I}_n = \{\boldsymbol{s}_0, \ldots, \boldsymbol{s}_n; \boldsymbol{u}_0, \ldots, \boldsymbol{u}_n\},$$

and the conditional mathematical expectation

$$\hat{\boldsymbol{x}}_n = \mathrm{E}[\boldsymbol{x}_n | \boldsymbol{I}_n]$$

is to be calculated. On account of Theorems 1–4, this problem reduces to projection of the $\mathrm{X}_n^{\,i}$ on the vector space $\mathscr{S}_n$ generated by the random variables $\boldsymbol{s}_0, \ldots, \boldsymbol{s}_n$ (i.e., the space spanned by the vectors $\mathscr{S}_k^{\,i}$ for $k = 0, \ldots, n$). This space can be decomposed into two parts:

$$\mathscr{S}_n = \mathscr{S}_{n-1} \oplus \mathscr{C}_n, \tag{A.5}$$

where $\mathscr{S}_{n-1}$ is the space corresponding to the information set $\boldsymbol{I}_{n-1}$, and $\mathscr{C}_n$ is the new part contributed by the knowledge of $\boldsymbol{s}_n$. Relation (A.5) simply indicates that each new piece of information augments the space $\mathscr{S}_n$.

We can now write

$$\boldsymbol{s}_n = \boldsymbol{s}_n^* + \tilde{\boldsymbol{s}}_n, \tag{A.6}$$

where

$$\boldsymbol{s}_n^* \in \mathscr{S}_{n-1}, \qquad \tilde{\boldsymbol{s}}_n \in \mathscr{C}_n.$$

The vector $\boldsymbol{s}_n^*$ is the projection of $\boldsymbol{s}_n$ on $\mathscr{S}_{n-1}$, and is thus the best estimate of $\boldsymbol{s}_n$ at time $n-1$ ($\boldsymbol{s}_n$ being a random variable). In the same way, we can write

$$\boldsymbol{x}_n = \boldsymbol{x}_n^* + \tilde{\boldsymbol{x}}_n. \tag{A.7}$$

The vector $\boldsymbol{x}_n^*$ is the best estimate of $\boldsymbol{x}_n$, conditioned by the information set $\boldsymbol{I}_{n-1}$, since it is the projection of $\boldsymbol{x}_n$ on $\mathscr{S}_{n-1}$. The vector $\boldsymbol{x}_{n+1}^*$ will thus be the best prediction of $\boldsymbol{x}_{n+1}$ at the time $n$.

We shall now derive the recursive filter defining $\boldsymbol{x}_n^*$ from which definition of $\hat{\boldsymbol{x}}_n$ follows easily.

### A.2.2 Derivation of the Recursive Predictor

In order to develop a recursive structure, let us write $\boldsymbol{x}_{n+1}^*$ as

$$\begin{aligned}\boldsymbol{x}_{n+1}^* &= (\text{projection of } \boldsymbol{x}_{n+1} \text{ on } \mathscr{S}_n)\\ &= (\text{projection of } \boldsymbol{x}_{n+1} \text{ on } \mathscr{S}_{n-1}) + (\text{projection of } \boldsymbol{x}_{n+1} \text{ on } \mathscr{C}_n).\end{aligned} \tag{A.8}$$

According to (A.1), the first projection, $\boldsymbol{x}_{n+1}$ on $\mathscr{S}_{n-1}$, is $\mathbf{F}_n \boldsymbol{x}_n^* + \mathbf{H}_n \boldsymbol{u}_n$, since $\boldsymbol{\delta}_n$ is independent of $\boldsymbol{s}_0, \ldots, \boldsymbol{s}_{n-1}$. The second projection belongs to the space $\mathscr{C}_n$ arising from $\boldsymbol{s}_n$ and will thus be of the form $\mathbf{K}_n \tilde{\boldsymbol{s}}_n$ or

$$\mathbf{K}_n(\boldsymbol{s}_n - \boldsymbol{s}_n^*).$$

This leads to

$$\boldsymbol{x}_{n+1}^* = \mathbf{F}_n \boldsymbol{x}_n^* + \mathbf{H}_n \boldsymbol{u}_n + \mathbf{K}_n(\boldsymbol{s}_n - \boldsymbol{s}_n^*). \tag{A.9}$$

The term $(\boldsymbol{s}_n - \boldsymbol{s}_n^*)$ is the difference between the predicted output and the output actually observed.

The term $\boldsymbol{s}_n^*$ can be obtained by projecting $\boldsymbol{s}_n$ on $\mathscr{S}_{n-1}$. Taking account of (A.1), the result is

$$\boldsymbol{s}_n^* = \mathbf{G}_n \boldsymbol{x}_n^*,$$

since $\boldsymbol{\delta}_n'$ is independent of the variables defining $\mathscr{S}_{n-1}$.

Thus the recursive equation for the predictor is

$$\boldsymbol{x}_{n+1}^* = \mathbf{F}_n \boldsymbol{x}_n^* + \mathbf{H}_n \boldsymbol{u}_n + \mathbf{K}_n(\boldsymbol{s}_n - \mathbf{G}_n \boldsymbol{x}_n^*). \tag{A.10}$$

The first part of (A.10) is analogous to the process equation, and the second is a correction term depending on the difference between the actual and predicted outputs.

### A.2.3 Calculation of the Predictor Coefficients

To calculate $\mathbf{K}_n$, we note that the vector

$$\boldsymbol{x}_{n+1} - (\text{projection of } \boldsymbol{x}_{n+1} \text{ on } \mathscr{C}_n)$$

is orthogonal to $\mathscr{C}_n$, and thus independent of $\tilde{\boldsymbol{s}}_n$. Also

$$\begin{aligned}\text{projection of } \boldsymbol{x}_{n+1} \text{ on } \mathscr{C}_n &= \text{projection of [projection of } \boldsymbol{x}_{n+1} \text{ on } \mathscr{S}_n] \text{ on } \mathscr{C}_n\\ &= \text{projection of } \boldsymbol{x}_{n+1}^* \text{ on } \mathscr{C}_n = \mathbf{K}_n \tilde{\boldsymbol{s}}_n,\end{aligned}$$

the first equality following because $\mathscr{C}_n \subset \mathscr{S}_n$. Hence

$$\mathrm{E}[\boldsymbol{x}_{n+1} \tilde{\boldsymbol{s}}_n^{\mathrm{T}}] - \mathbf{K}_n \mathrm{E}[\tilde{\boldsymbol{s}}_n \tilde{\boldsymbol{s}}_n^{\mathrm{T}}] = \mathbf{0},$$

from which

$$\mathbf{K}_n = \mathrm{E}[\boldsymbol{x}_{n+1}\tilde{\boldsymbol{s}}_n^{\mathrm{T}}]\{\mathrm{E}[\tilde{\boldsymbol{s}}_n\tilde{\boldsymbol{s}}_n^{\mathrm{T}}]\}^{-1}. \tag{A.11}$$

Thus calculation of $\mathbf{K}_n$ requires calculation of two covariance matrices:

$$\begin{aligned}\mathrm{E}[\boldsymbol{x}_{n+1}\tilde{\boldsymbol{s}}_n^{\mathrm{T}}] &= \mathrm{E}[(\mathbf{F}_n(\boldsymbol{x}_n^* + \tilde{\boldsymbol{x}}_n) + \mathbf{H}_n\boldsymbol{u}_n + \boldsymbol{\delta}_n)(\mathbf{G}_n\tilde{\boldsymbol{x}}_n + \boldsymbol{\delta}_n')^{\mathrm{T}}] \\ &= \mathbf{F}_n\mathrm{E}[\tilde{\boldsymbol{x}}_n\tilde{\boldsymbol{x}}_n^{\mathrm{T}}]\mathbf{G}_n^{\mathrm{T}},\end{aligned} \tag{A.12}$$

$$\mathrm{E}[\tilde{\boldsymbol{s}}_n\tilde{\boldsymbol{s}}_n^{\mathrm{T}}] = \mathrm{E}[(\mathbf{G}_n\tilde{\boldsymbol{x}}_n + \boldsymbol{\delta}_n')(\mathbf{G}_n\tilde{\boldsymbol{x}}_n + \boldsymbol{\delta}_n')^{\mathrm{T}}] = \mathbf{G}_n\mathrm{E}[\tilde{\boldsymbol{x}}_n\tilde{\boldsymbol{x}}_n^{\mathrm{T}}]\mathbf{G}_n^{\mathrm{T}} + \boldsymbol{\Phi}_n'. \tag{A.13}$$

Letting

$$\boldsymbol{\Psi}_n = \mathrm{E}[\tilde{\boldsymbol{x}}_n\tilde{\boldsymbol{x}}_n^{\mathrm{T}}], \tag{A.14}$$

which is the covariance matrix of the prediction error, there results from (A.12), (A.13), and (A.14),

$$\mathbf{K}_n = \mathbf{F}_n\boldsymbol{\Psi}_n\mathbf{G}_n^{\mathrm{T}}[\mathbf{G}_n\boldsymbol{\Psi}_n\mathbf{G}_n^{\mathrm{T}} + \boldsymbol{\Phi}_n']^{-1}. \tag{A.15}$$

It remains only to calculate $\boldsymbol{\Psi}_n$. For this we again seek a recursive solution using the recurrence relation for $\tilde{\boldsymbol{x}}_n$:

$$\begin{aligned}\tilde{\boldsymbol{x}}_{n+1} &= \boldsymbol{x}_{n+1} - \boldsymbol{x}_{n+1}^* \\ &= \mathbf{F}_n\boldsymbol{x}_n + \mathbf{H}_n\boldsymbol{u}_n + \boldsymbol{\delta}_n - [\mathbf{F}_n\boldsymbol{x}_n^* + \mathbf{H}_n\boldsymbol{u}_n + \mathbf{K}_n(\mathbf{G}_n\boldsymbol{x}_n + \boldsymbol{\delta}_n' - \mathbf{G}_n\boldsymbol{x}_n^*)] \\ &= (\mathbf{F}_n - \mathbf{K}_n\mathbf{G}_n)\tilde{\boldsymbol{x}}_n + \boldsymbol{\delta}_n - \mathbf{K}_n\boldsymbol{\delta}_n'.\end{aligned} \tag{A.16}$$

Since $\boldsymbol{\delta}_n$ and $\boldsymbol{\delta}_n'$ are independent, it follows from (A.16) that

$$\boldsymbol{\Psi}_{n+1} = (\mathbf{F}_n - \mathbf{K}_n\mathbf{G}_n)\boldsymbol{\Psi}_n(\mathbf{F}_n - \mathbf{K}_n\mathbf{G}_n)^{\mathrm{T}} + \boldsymbol{\Phi}_n + \mathbf{K}_n\boldsymbol{\Phi}_n'\mathbf{K}_n^{\mathrm{T}}. \tag{A.17}$$

Equations (A.10), (A.15), and (A.17) form the solution to the optimal prediction problem.

If the measurement noise is zero, $\boldsymbol{\Phi}_n' = \mathbf{0}$, and the relation (A.17) simplifies to

$$\boldsymbol{\Psi}_{n+1} = (\mathbf{F}_n - \mathbf{K}_n\mathbf{G}_n)\boldsymbol{\Psi}_n\mathbf{F}_n^{\mathrm{T}} + \boldsymbol{\Phi}_n.$$

### *A.2.4 The Optimal Estimator*

The estimate $\hat{\boldsymbol{x}}_n$ is the projection of $\boldsymbol{x}_n$ on $\mathscr{S}_n$. If $\boldsymbol{x}_{n+1}$ is projected on $\mathscr{S}_n$, there results from the definitions and relation (A.1):

$$\boldsymbol{x}_{n+1}^* = \mathbf{F}_n\hat{\boldsymbol{x}}_n + \mathbf{H}_n\boldsymbol{u}_n. \tag{A.18}$$

To obtain a recurrence relation for $\hat{\boldsymbol{x}}_n$, it is only necessary to substitute (A.18) into (A.10), to obtain

$$\mathbf{F}_n \hat{\boldsymbol{x}}_n + \mathbf{H}_n \boldsymbol{u}_n = \mathbf{F}_n(\mathbf{F}_{n-1}\hat{\boldsymbol{x}}_{n-1} + \mathbf{H}_{n-1}\boldsymbol{u}_{n-1}) + \mathbf{H}_n \boldsymbol{u}_n + \mathbf{K}_n[\boldsymbol{s}_n - \mathbf{G}_n(\mathbf{F}_{n-1}\hat{\boldsymbol{x}}_{n-1} + \mathbf{H}_{n-1}\boldsymbol{u}_{n-1})].$$

Assuming that $\mathbf{F}_n^{-1}$ exists, and defining

$$\mathbf{M}_n = \mathbf{F}_n^{-1}\mathbf{K}_n,$$

this becomes

$$\hat{\boldsymbol{x}}_n = \mathbf{F}_{n-1}\hat{\boldsymbol{x}}_{n-1} + \mathbf{H}_{n-1}\boldsymbol{u}_{n-1} + \mathbf{M}_n[\boldsymbol{s}_n - \mathbf{G}_n(\mathbf{F}_{n-1}\hat{\boldsymbol{x}}_{n-1} + \mathbf{H}_{n-1}\boldsymbol{u}_{n-1})]. \tag{A.19}$$

This is the recursive equation for the optimal estimator.

From (A.15), it follows that $\mathbf{M}_n$ is given by

$$\mathbf{M}_n = \boldsymbol{\Psi}_n \mathbf{G}_n^{\mathrm{T}}[\mathbf{G}_n \boldsymbol{\Psi}_n \mathbf{G}_n^{\mathrm{T}} + \boldsymbol{\Phi}_n']^{-1}. \tag{A.20}$$

Replacing $\mathbf{K}_n$ in (A.17) by $\mathbf{F}_n \mathbf{M}_n$ yields

$$\boldsymbol{\Psi}_{n+1} = \mathbf{F}_n(\mathbf{1} - \mathbf{M}_n \mathbf{G}_n)\boldsymbol{\Psi}_n(\mathbf{1} - \mathbf{M}_n \mathbf{G}_n)^{\mathrm{T}}\mathbf{F}_n^{\mathrm{T}} + \boldsymbol{\Phi}_n + \mathbf{F}_n \mathbf{M}_n \boldsymbol{\Phi}_n' \mathbf{M}_n^{\mathrm{T}} \mathbf{F}_n^{\mathrm{T}}. \tag{A.21}$$

### *A.2.5 Implementation*

Fig. A.2 shows the block diagram of the optimal predictor, described by equations (A.10), (A.15), and (A.17), and Fig. A.3 shows the optimal estimator, described by (A.19), (A.20), and (A.21). Note that all calculations of $\boldsymbol{\Psi}_n$ and $\mathbf{K}_n$ or $\mathbf{M}_n$ are carried out in forward time. They can thus be done in real time and remain valid for any sampling interval, provided that at each cycle the appropriate $\mathbf{F}_n$, $\mathbf{G}_n$, and $\mathbf{H}_n$ are used. In particular, this applies to prediction over an interval longer than one sample period when the $\boldsymbol{u}_n$ are zero or constant.

If $\boldsymbol{x}_0$ is perfectly known, its value is used to initialize equations (A.10) or (A.19), and $\boldsymbol{\Psi}_0 = \mathbf{0}$ is used in (A.17) or (A.21). If $\boldsymbol{x}_0$ is a Gaussian random variable with mean $\bar{\boldsymbol{x}}_0$ and covariance $\boldsymbol{\Phi}_{x_0x_0}$, the appropriate initial conditions are

$$\boldsymbol{x}_0{}^* = \bar{\boldsymbol{x}}_0 \quad \text{or} \quad \hat{\boldsymbol{x}}_0 = \bar{\boldsymbol{x}}_0,$$

$$\boldsymbol{\Psi}_0 = \boldsymbol{\Phi}_{x_0x_0}.$$

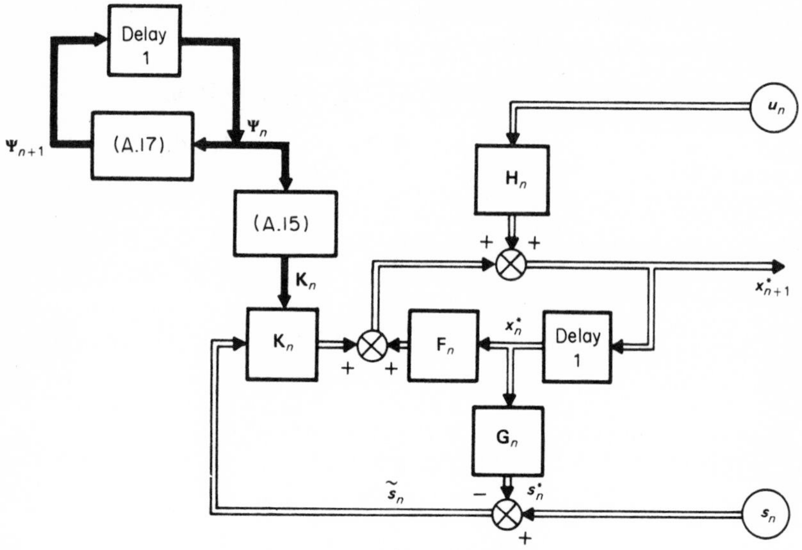


FIG. A.2. The optimal discrete predictor.

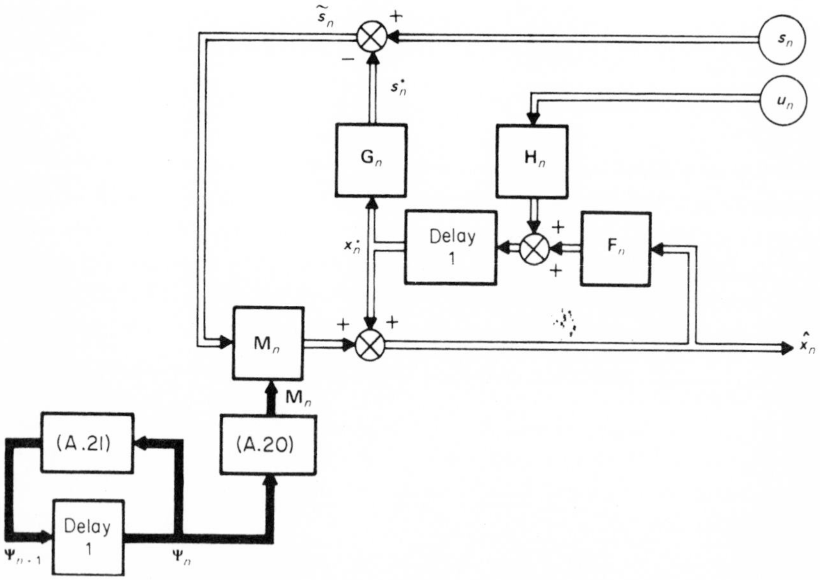


FIG. A.3. The optimal discrete estimator.

### A.2.6 *Example*

Consider the problem indicated in Fig. A.4, corresponding to the state vector model

$$x^+ = \begin{bmatrix} -0.5 & 0 \\ 0.25 & -0.25 \end{bmatrix} x + \begin{bmatrix} 0.5 \\ 0.125 \end{bmatrix} b, \qquad \varphi_b = 64,$$

$$s = [0 \quad 1] x + v \qquad \varphi_v = 1.$$

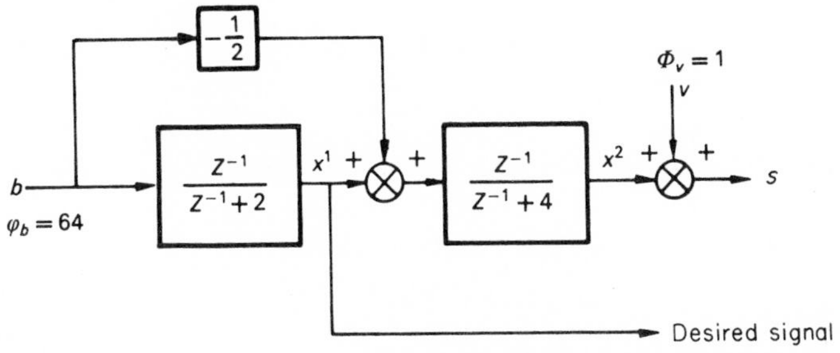


FIG. A.4. A noisy system with a state variable to be estimated.

The variable to be estimated is $x_n^1$. The perturbation covariance matrices are

$$\mathbf{\Phi}_n = \begin{bmatrix} 16 & -4 \\ -4 & 1 \end{bmatrix}, \qquad \varphi_n' = 1.$$

Letting

$$\mathbf{\Psi}_n = \begin{bmatrix} a & b \\ b & c \end{bmatrix},$$

there results

$$\mathbf{M}_n = \begin{bmatrix} a & b \\ b & c \end{bmatrix} \begin{bmatrix} 0 \\ 1 \end{bmatrix} \left\{ [0 \quad 1] \begin{bmatrix} a & b \\ b & c \end{bmatrix} \begin{bmatrix} 0 \\ 1 \end{bmatrix} + 1 \right\}^{-1}$$

$$= \frac{1}{c+1} \begin{bmatrix} b \\ c \end{bmatrix}.$$

We shall not pursue this further, and will not calculate (A.21) in expanded literal form. In fact, it is preferable to calculate (A.21) in the sequence of steps

$$\mathbf{A} \Leftarrow \mathbf{MG},$$
$$\mathbf{A} \Leftarrow \mathbf{1} - \mathbf{A},$$
$$\mathbf{B} \Leftarrow \mathbf{FA},$$
$$\mathbf{\Psi}^+ \Leftarrow \mathbf{B\Psi B}^{\mathrm{T}},$$
$$\mathbf{\Psi}^+ \Leftarrow \mathbf{\Psi}^+ + \mathbf{\Phi},$$
$$\mathbf{A} \Leftarrow \mathbf{FM},$$
$$\mathbf{B} \Leftarrow \mathbf{A\Phi}'\mathbf{A}^{\mathrm{T}},$$
$$\mathbf{\Psi}^+ \Leftarrow \mathbf{\Psi}^+ + \mathbf{B}.$$

## A.3 Estimation of the State of a Continuous Process

Let us now consider the continuous linear process

$$\begin{aligned} \mathrm{d}\boldsymbol{x} &= (\mathbf{A}\boldsymbol{x} + \mathbf{B}\boldsymbol{u})\,\mathrm{d}t + \mathrm{d}\boldsymbol{\beta}, \\ \boldsymbol{s} &= \mathbf{C}\boldsymbol{x} + \mathrm{d}\boldsymbol{\beta}'/\mathrm{d}t, \end{aligned} \tag{A.22}$$

where $\boldsymbol{\beta}$ and $\boldsymbol{\beta}'$ are Gaussian random processes with independent increments, such that

$$\lim_{\mathrm{d}t \to 0} \frac{1}{\mathrm{d}t} \mathrm{E}[\mathrm{d}\boldsymbol{\beta}] = \mathbf{0}, \qquad \lim_{\mathrm{d}t \to 0} \frac{1}{\mathrm{d}t} \mathrm{E}[\mathrm{d}\boldsymbol{\beta}\,\mathrm{d}\boldsymbol{\beta}^{\mathrm{T}}] = \mathbf{\Pi}, \tag{A.23}$$

$$\lim_{\mathrm{d}t \to 0} \frac{1}{\mathrm{d}t} \mathrm{E}[\mathrm{d}\boldsymbol{\beta}'] = \mathbf{0}, \qquad \lim_{\mathrm{d}t \to 0} \frac{1}{\mathrm{d}t} \mathrm{E}[\mathrm{d}\boldsymbol{\beta}'\,\mathrm{d}\boldsymbol{\beta}'^{\mathrm{T}}] = \mathbf{\Pi}'. \tag{A.24}$$

The processes $\mathrm{d}\boldsymbol{\beta}/\mathrm{d}t$ and $\mathrm{d}\boldsymbol{\beta}'/\mathrm{d}t$ are thus " white " noise processes, with autocorrelations $\mathbf{\Pi}$ and $\mathbf{\Pi}'$.

### *A.3.1 The Optimal Filter*

We shall consider (A.22) as the limit of the system of difference equations

$$\Delta\boldsymbol{x} = (\mathbf{A}\boldsymbol{x} + \mathbf{B}\boldsymbol{u})T + \boldsymbol{\delta}_T, \tag{A.25}$$

$$\boldsymbol{s} = \mathbf{C}\boldsymbol{x} + \boldsymbol{\delta}_T'/T, \tag{A.26}$$

in which the perturbations $\boldsymbol{\delta}_T$ and $\boldsymbol{\delta}_T'$ are centered, independent of each other, such that

$$\lim_{T\to 0} \frac{1}{T} \mathrm{E}[\boldsymbol{\delta}_T \, \boldsymbol{\delta}_T{}^{\mathrm{T}}] = \boldsymbol{\Pi}, \tag{A.27}$$

$$\lim_{T\to 0} \frac{1}{T} \mathrm{E}[\boldsymbol{\delta}_T' \, \boldsymbol{\delta}_T'^{\mathrm{T}}] = \boldsymbol{\Pi}'. \tag{A.28}$$

The system (A.25) and (A.26) can then be put in the form of (A.1), as

$$\begin{aligned} \boldsymbol{x}^+ &= (\mathbf{1} + T\mathbf{A})\boldsymbol{x} + T\mathbf{B}\boldsymbol{u} + \boldsymbol{\delta}, \\ \boldsymbol{s} &= \mathbf{C}\boldsymbol{x} + \boldsymbol{\delta}'/T. \end{aligned} \tag{A.29}$$

Thus we have the correspondences

$$\mathbf{F} = \mathbf{1} + T\mathbf{A}, \qquad \mathbf{H} = T\mathbf{B}, \qquad \mathbf{G} = \mathbf{C},$$

and

$$\boldsymbol{\Phi} \simeq \boldsymbol{\Pi} T, \qquad \boldsymbol{\Phi}' \simeq \boldsymbol{\Pi}'/T.$$

When $T \to 0$, the distinction between prediction and estimation disappears, and we may use formulas (A.10), (A.15), and (A.17) to determine the continuous estimator.

Formula (A.10) can be written

$$\boldsymbol{x}^{*+} = (\mathbf{1} + T\mathbf{A})\boldsymbol{x}^* + T\mathbf{B}\boldsymbol{u} + \mathbf{K}(\boldsymbol{s} - \mathbf{C}\boldsymbol{x}^*),$$

or, after introducing $\Delta\boldsymbol{x}^* = \boldsymbol{x}^{*+} - \boldsymbol{x}^*$ and letting $T$ tend to zero,

$$\dot{\boldsymbol{x}}^* = \mathbf{A}\boldsymbol{x}^* + \mathbf{B}\boldsymbol{u} + \boldsymbol{\Lambda}(\boldsymbol{s} - \mathbf{C}\boldsymbol{x}^*), \tag{A.30}$$

where

$$\boldsymbol{\Lambda} = \mathbf{K}/T.$$

The structure of this filter, indicated in Fig. A.5, is analogous to that in the discrete case.

Relation (A.15) becomes

$$T\boldsymbol{\Lambda} = (\mathbf{1} + T\mathbf{A})\boldsymbol{\Psi}\mathbf{C}^{\mathrm{T}}(\mathbf{C}\boldsymbol{\Psi}\mathbf{C}^{\mathrm{T}} + \boldsymbol{\Pi}'/T)^{-1},$$

or, for $T \to 0$,

$$\boldsymbol{\Lambda} = \boldsymbol{\Psi}\mathbf{C}^{\mathrm{T}}\boldsymbol{\Pi}'^{-1}. \tag{A.31}$$

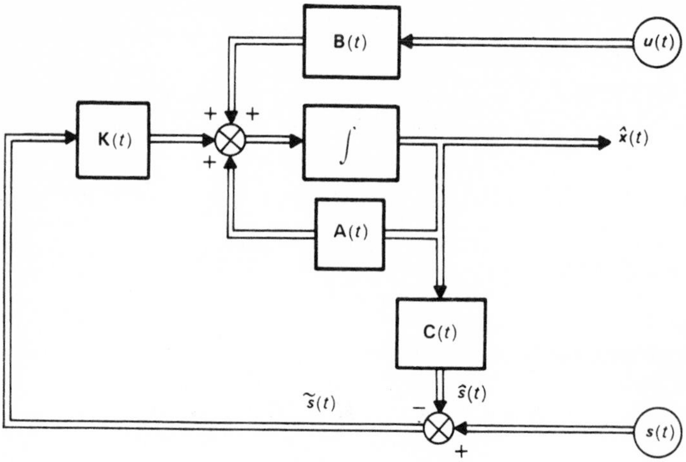


FIG. A.5. Optimal continuous estimator.

Taking account of (A.31), relation (A.17) becomes

$$\mathbf{\Psi}^{+} = (\mathbf{1} + T\mathbf{A} - T\mathbf{\Psi}\mathbf{C}^{\mathrm{T}}\mathbf{\Pi}'^{-1}\mathbf{C})\mathbf{\Psi}(\mathbf{1} + T\mathbf{A} - T\mathbf{\Psi}\mathbf{C}^{\mathrm{T}}\mathbf{\Pi}'^{-1}\mathbf{C})^{\mathrm{T}} + \mathbf{\Pi}T + T\mathbf{\Psi}\mathbf{C}^{\mathrm{T}}\mathbf{\Pi}'^{-1}(\mathbf{\Pi}'/T)\mathbf{\Pi}'^{-1}\mathbf{C}\mathbf{\Psi}T,$$

or, after passage to the limit and simplification,

$$\dot{\mathbf{\Psi}} = \mathbf{A}\mathbf{\Psi} + \mathbf{\Psi}\mathbf{A}^{\mathrm{T}} - \mathbf{\Psi}\mathbf{C}^{\mathrm{T}}\mathbf{\Pi}'^{-1}\mathbf{C}\mathbf{\Psi} + \mathbf{\Pi}. \tag{A.32}$$

This is a nonlinear matrix equation of the Riccati type.

Note that the above equations make sense only if $\mathbf{\Pi}'$ is invertible. This will be the case so long as all components of the measurement noise $\boldsymbol{\beta}'$ are nonzero and statistically independent. When this is not the case, the modified formulas established by Kalman and Bucy can be used.

### *A.3.2 Practical Implementation*

Solution of (A.32) is difficult, and in general must be carried out numerically, since only in very simple cases is a literal solution possible.

Operation of (A.30) requires synthesis of variable gains, since even if $\mathbf{A}(t)$, $\mathbf{B}(t)$, and $\mathbf{C}(t)$ are stationary, $\mathbf{K}(t)$ is in general variable. This complication, in most cases, requires realization of the filter using a real-time digital computer. Thus, even for a problem of only moderate complexity, calculation of $\mathbf{K}(t)$ and realization of the filter require numerical methods, and thus discretization of the appropriate equations. It would then appear that discretization of the process before design of the filter might be preferable, in which case only the results of Section A.2 are needed. Choice of the sampling period in any such discretization should take into account the spectrum of the measurement noise in order to avoid any frequency aliasing.

## A.4 Conclusions

In this appendix, we have studied the Kalman filter. In fact, any problem of signal filtering, in which the signals are specified in terms of the autocorrelations of $\mathbf{s}(t)$ and $\mathbf{d}(t)$ and their cross correlations, can be reduced to estimation of the state of a generator of the process [1, Vol. 1, Part 3], so that the method treated here is quite general.

We have emphasized the importance of the discrete formalism, which leads to an easily computed and implemented solution. This requires preliminary discretization of the problem, if the problem is initially continuous [1, Vol. 1, Ch. 14].

We have not touched upon the question of the stability of the Kalman filter. This involves study of the behavior of $\mathbf{\Phi}_n$ as $n \to \infty$. Usually the filter tends to a stable stationary filter, identical to the filter which would be obtained using the methods of Wiener, which always assume an infinite horizon.

Another important question is that of the sensitivity of the filter. This arises in cases in which the parameters of the process are either not precisely known, or variable. The question is of particular interest if the process is unstable.

# References


1. R. Boudarel, J. Delmas, and P. Guichet, "Commande Optimale des Processus." Vols. 1, 2, and 4, Dunod, Paris, 1967, 1968, and 1970.
2. R. Bellman, "Dynamic Programming." Princeton Univ. Press, Princeton, New Jersey, 1957.
3. R. Bellman, "Adaptive Control Processes: A Guided Tour." Princeton Univ. Press, Princeton, New Jersey, 1961.
4. R. Bellman and S. Dreyfus, "Applied Dynamic Programming." Princeton Univ. Press, Princeton, New Jersey, 1962.
5. S. Roberts, "Dynamic Programming in Chemical Engineering and Process Control." Academic Press, New York, 1964.
6. C. W. Merriam, III, "Optimization Theory and the Design of Feedback Control Systems." McGraw-Hill, New York, 1964.
7. J. Tou, "Optimum Design of Digital Control Systems." Academic Press, New York, 1963.
8. R. Bellman, Dynamic programming and stochastic control processes, *Information and Control*, **1**, 228–239 (1958).
9. R. Bellman, On the application of the theory of dynamic programming to the study of control processes, *Proc. Symp. on Non-Linear Circuit Analysis*, Polytechnic Institute of Brooklyn, Brooklyn, New York, 1957.
10. R. Bellman and R. Kalaba, Dynamic programming and adaptive processes.—I: Mathematical foundation, *IRE Trans. Auto. Control*, **AC-5** (1), 5–10 (1960).
11. R. Bellman and R. Kalaba, On adaptive control processes, *IRE Trans. Auto. Control*, **AC-4**, 1–9 (1959).
12. R. Bellman, J. Holland, and R. Kalaba, On the application of dynamic programming to the synthesis of logical systems, *J. Assoc. Comput. Mach.*, **6**, 486–493 (1959).
13. R. Bellman and R. Kalaba, Dynamic programming and feedback control, (First IFAC Congress, Moscow, 1960, *Automat. Remote Contr.* (Butterworth, London) **1**, 460–464 (1961).

14. R. Bellman, On the determination of optimal trajectories via dynamic programming, *in* "Optimization Techniques: With Applications to Aerospace Systems" (G. Leitmann, ed.). Academic Press, New York, 1962.
15. M. Aoki, Dynamic Programming and Numerical Experimentation as Applied to Adaptive Control Systems. Ph. D. Thesis, Univ. of California (Los Angeles), Los Angeles, California, 1959.
16. E. Peterson, Optimal Control for Stochastic and Adaptive Processes. Report RM 61 TMP-73, General Electric Company, Santa Barbara, California, 1961.
17. P. Joseph, Multivariable linear optimal control, *Proc. Symp. Multivariable Systems, Massachusetts Institute of Technology, November 1962.* M.I.T. Press, Cambridge, Massachusetts.
18. T. Gunckel, Optimum design of sampled data systems with random parameters, *Proc. Symp. Multivariable Systems, Massachusetts Institute of Technology, Cambridge, November 1962.* M.I.T. Press, Cambridge, Massachusetts.
19. C. W. Merriam, III, Use of a mathematical error criterion in the design of adaptive control systems, *AIEE Trans.*, Part II, **78**, 506–512 (1959).
20. P. Joseph and J. Tou, On linear control theory. *AIEE Trans.* Part II, **80**, 193–196 (1961).
21. J. H. Eaton and L. A. Zadeh, Optimal pursuit strategies in discrete state probabilistic systems, *J. Basic Eng.* **84**, 23–29 (1962).
22. J. Guignabodet, Some bounds on quantization errors in dynamic programming computations, *Proc. Second IFAC Congress, Basel, 1963*, 383–385. Butterworth, London, 1964.
23. P. Guichet and R. Boudarel, Synthese de la commande numerique optimale des processus lineaires, *Ann. Radioelec.*, **18**, 291–304 (1963).
24. R. Bellman and R. Kalaba, Dynamic programming, invariant imbedding and quasilinearization—Comparisons and Interconnections, *in* "Computing Methods in Optimization Problems," (A. Balakrishnan and L. Neustadt, eds.). Academic Press, New York, 1964.
25. R. Kalman, A new approach to linear filtering and prediction problems, *J. Basic Eng.*, **82**, 35–44 (1960).
26. Y. C. Ho, The Method of Least Squares and Optimal Filtering Theory. RAND Corporation Memorandum RM-3329-PR, October, 1962.
27. S. Fagin, Recursive linear regression theory, optimal filter theory, and error analyses of optimal systems, *IEEE Internat. Conv. Rec.*, Part 1, 216–240 (1964).
28. E. Parzen, "Modern Probability Theory and Its Applications." Wiley, New York, 1960.
29. L. Liusternick and V. Sobolev, "Elements of Functional Analysis." Ungar, New York, 1961.
30. R. Kalman and J. Bertram, Control system analysis and design via the second method of Lyapunov: II—Discrete time systems, *J. Basic Eng.* **82**, 394–400 (1960).
31. A. Bastiani, Systémes guidables et problèmes d'optimisation, *Proc. Conf. Theory of Automation, Saclay, 1965.*
32. P. Loriquet, Quadratic optimization of linear discrete systems, *Proc. IFAC Symposium on Multivariable Regulators. October 1968.*

# Index

# Mathematics in Science and Engineering

## *A Series of Monographs and Textbooks*

**Edited by RICHARD BELLMAN,** *University of Southern California*

**23.** A. Halanay. Differential Equations: Stability, Oscillations, Time Lags. 1966

**24.** M. N. Oguztoreli. Time-Lag Control Systems. 1966

**25.** D. Sworder. Optimal Adaptive Control Systems. 1966

**26.** M. Ash. Optimal Shutdown Control of Nuclear Reactors. 1966

**27.** D. N. Chorafas. Control System Functions and Programming Approaches (In Two Volumes). 1966

**28.** N. P. Erugin. Linear Systems of Ordinary Differential Equations. 1966

**29.** S. Marcus. Algebraic Linguistics; Analytical Models. 1967

**30.** A. M. Liapunov. Stability of Motion. 1966

**31.** G. Leitmann (ed.). Topics in Optimization. 1967

**32.** M. Aoki. Optimization of Stochastic Systems. 1967

**33.** H. J. Kushner. Stochastic Stability and control. 1967

**34.** M. Urabe. Nonlinear Autonomous Oscillations. 1967

**35.** F. Calogero. Variable Phase Approach to Potential Scattering. 1967

**36.** A. Kaufmann. Graphs, Dynamic Programming, and Finite Games. 1967

**37.** A. Kaufmann and R. Cruon. Dynamic Programming: Sequential Scientific Management. 1967

**38.** J. H. Ahlberg, E. N. Nilson, and J. L. Walsh. The Theory of Splines and Their Applications. 1967

**39.** Y. Sawaragi, Y. Sunahara, and T. Nakamizo. Statistical Decision Theory in Adaptive Control Systems. 1967

**40.** R. Bellman. Introduction to the Mathematical Theory of Control Processes, Volume I. 1967; Volume II. 1971 (Volume III in preparation)

**41.** E. S. Lee. Quasilinearization and Invariant Imbedding. 1968

**42.** W. Ames. Nonlinear Ordinary Differential Equations in Transport Processes. 1968

**43.** W. Miller, Jr. Lie Theory and Special Functions. 1968

**44.** P. B. Bailey, L. F. Shampine, and P. E. Waltman. Nonlinear Two Point Boundary Value Problems. 1968

**45.** Iu. P. Petrov. Variational Methods in Optimum Control Theory. 1968

**46.** O. A. Ladyzhenskaya and N. N. Ural't-seva. Linear and Quasilinear Elliptic Equations. 1968

**47.** A. Kaufmann and R. Faure. Introduction to Operations Research. 1968

**48.** C. A. Swanson. Comparison and Oscillation Theory of Linear Differential Equations. 1968

**49.** R. Hermann. Differential Geometry and the Calculus of Variations. 1968

**50.** N. K. Jaiswal. Priority Queues. 1968

**51.** H. Nikaido. Convex Structures and Economic Theory. 1968

**52.** K. S. Fu. Sequential Methods in Pattern Recognition and Machine Learning. 1968

**53.** Y. L. Luke. The Special Functions and Their Approximations (In Two Volumes). 1969

**54.** R. P. Gilbert. Function Theoretic Methods in Partial Differential Equations. 1969

**55.** V. Lakshmikantham and S. Leela. Differential and Integral Inequalities (In Two Volumes). 1969

**56.** S. H. Hermes and J. P. LaSalle. Functional Analysis and Time Optimal Control. 1969

**57.** M. Iri. Network Flow, Transportation, and Scheduling: Theory and Algorithms. 1969

**58.** A. Blaquiere, F. Gerard, and G. Leitmann. Quantitative and Qualitative Games. 1969

**59.** P. L. Falb and J. L. de Jong. Successive Approximation Methods in Control and Oscillation Theory. 1969

**60.** G. Rosen. Formulations of Classical and Quantum Dynamical Theory. 1969

**61.** R. Bellman. Methods of Nonlinear Analysis, Volume I. 1970

**62.** R. Bellman, K. L. Cooke, and J. A. Lockett. Algorithms, Graphs, and Computers. 1970

**63.** E. J. Beltrami. An Algorithmic Approach to Nonlinear Analysis and Optimization. 1970

**64.** A. H. Jazwinski. Stochastic Processes and Filtering Theory. 1970

**65.** P. Dyer and S. R. McReynolds. The Computation and Theory of Optimal Control. 1970

**66.** J. M. Mendel and K. S. Fu (eds.). Adaptive, Learning, and Pattern Recognition Systems: Theory and Applications. 1970

**67.** C. Derman. Finite State Markovian Decision Processes. 1970

**68.** M. Mesarovic, D. Macko, and Y. Takahara. Theory of Hierarchial Multilevel Systems. 1970

**69.** H. H. Happ. Diakoptics and Networks. 1971

**70.** Karl Astrom. Introduction to Stochastic Control Theory. 1970

**71.** G. A. Baker, Jr. and J. L. Gammel (eds.). The Padé Approximant in Theoretical Physics. 1970

**72.** C. Berge. Principles of Combinatorics. 1971

**73.** Ya. Z. Tsypkin. Adaptation and Learning in Automatic Systems. 1971

**74.** Leon Lapidus and John H. Seinfeld. Numerical Solution of Ordinary Differential Equations. 1971

**75.** L. Mirsky. Transversal Theory, 1971

**76.** Harold Greenberg. Integer Programming, 1971

**77.** E. Polak. Computational Methods in Optimization: A Unified Approach, 1971

**78.** Thomas G. Windeknecht. General Dynamical Processes: A Mathematical Introduction, 1971

**79.** M. A. Aiserman, L. A. Gusev, L. I. Rozonoer, I. M. Smirnova, and A. A. Tal'. Logic, Automata, and Algorithms, 1971

**80.** Andrew P. Sage and James L. Melsa. System Identification, 1971

**81.** R. Boudarel, J. Delmas, and P. Guichet. Dynamic Programming and Its Application to Optimal Control, 1971

*In preparation*

Alexander Weinstein and William Stenger. Methods of Intermediate Problems for Eigenvalues Theory and Ramifications

G. Arthur Mihram. Simulation: Statistical Foundations and Methodology

William S. Meisel. Computer-Oriented Approaches to Pattern Recognition

Edward Angel and Richard Bellman. Dynamic Programming and Partial Differential Equations

F. V. Atkinson. Multiparameter Eigenvalue Problems, Volume I

Bruce A. Finlayson. The Method of Weighted Residuals and Variational Principles: With Application to Fluid Mechanics, Heat and Mass Transfer